Tucholsky Wagner Zola Scott Sydow Freud Schlegel
Turgenev Fonatne Wallace
Twain Walther von der Vogelweide Fouqué Friedrich II. von Preußen
Weber Freiligrath Frey
Kant Ernst
Fechner Fichte Weiße Rose von Fallersleben Richthofen Frommel
Hölderlin
Engels Fielding Eichendorff Tacitus Dumas
Fehrs Faber Flaubert
Eliasberg Ebner Eschenbach
Feuerbach Maximilian I. von Habsburg Fock Zweig
Eliot
Ewald Vergil
Goethe Elisabeth von Österreich London
Mendelssohn Balzac Shakespeare Dostojewski Ganghofer
Lichtenberg Rathenau Doyle Gjellerup
Trackl Stevenson Hambruch
Mommsen Tolstoi Lenz Droste-Hülshoff
Thoma Hanrieder
Dach Verne von Arnim Hägele Hauff Humboldt
Reuter Rousseau Hagen Hauptmann Gautier
Karrillon Garschin
Defoe Hebbel Baudelaire
Damaschke Descartes
Hegel Kussmaul Herder
Wolfram von Eschenbach Dickens Schopenhauer Rilke George
Bronner Darwin Melville Grimm Jerome
Campe Horváth Aristoteles Bebel Proust
Bismarck Vigny Barlach Voltaire Federer Herodot
Gengenbach Heine
Storm Casanova Tersteegen Grillparzer Georgy
Chamberlain Lessing Langbein Gilm Gryphius
Brentano Lafontaine
Strachwitz Claudius Schiller Kralik Iffland Sokrates
Katharina II. von Rußland Bellamy Schilling
Gerstäcker Raabe Gibbon Tschechow
Löns Hesse Hoffmann Gogol Wilde Vulpius
Luther Heym Hofmannsthal Gleim
Roth Klee Hölty Morgenstern
Luxemburg Heyse Klopstock Kleist Goedicke
Machiavelli La Roche Puschkin Homer Mörike Musil
Horaz
Navarra Aurel Musset Kierkegaard Kraft Kraus
Nestroy Marie de France Lamprecht Kind Kirchhoff Hugo Moltke
Laotse Ipsen Liebknecht
Nietzsche Nansen Ringelnatz
Marx Lassalle Gorki Klett Leibniz
von Ossietzky May vom Stein Lawrence Irving
Petalozzi Platon Knigge
Sachs Pückler Michelangelo Kock Kafka
Poe Liebermann
de Sade Praetorius Mistral Zetkin Korolenko

The publishing house tredition has created the series **TREDITION CLASSICS**. It contains classical literature works from over two thousand years. Most of these titles have been out of print and off the bookstore shelves for decades.

The book series is intended to preserve the cultural legacy and to promote the timeless works of classical literature. As a reader of a **TREDITION CLASSICS** book, the reader supports the mission to save many of the amazing works of world literature from oblivion.

The symbol of **TREDITION CLASSICS** is Johannes Gutenberg (1400 – 1468), the inventor of movable type printing.

With the series, tredition intends to make thousands of international literature classics available in printed format again – worldwide.

All books are available at book retailers worldwide in paperback and in hardcover. For more information please visit: www.tredition.com

tredition was established in 2006 by Sandra Latusseck and Soenke Schulz. Based in Hamburg, Germany, tredition offers publishing solutions to authors and publishing houses, combined with worldwide distribution of printed and digital book content. tredition is uniquely positioned to enable authors and publishing houses to create books on their own terms and without conventional manufacturing risks.

For more information please visit: www.tredition.com

Carpentry for Boys In a Simple Language, Including Chapters on Drawing, Laying Out Work, Designing and Architecture With 250 Original Illustrations

James Slough Zerbe

Imprint

This book is part of the TREDITION CLASSICS series.

Author: James Slough Zerbe
Cover design: toepferschumann, Berlin (Germany)

Publisher: tredition GmbH, Hamburg (Germany)
ISBN: 978-3-8491-5344-1

www.tredition.com
www.tredition.de

The "How-to-do-it" Books

CARPENTRY FOR BOYS

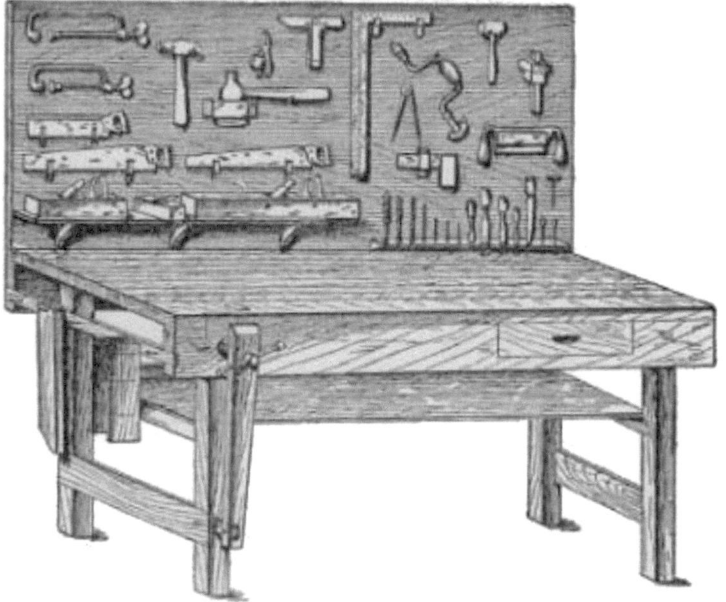

Fig. 1. A Typical Work Bench.

THE NEW YORK BOOK COMPANY

[Pg i]

CONTENTS

Knowledge of Tools. A Full Kit of Tools. The Hatchet. The Claw Hammer. About Saws—Cross-cut, Rip Saw, Back Saw. Planes—Jack Plane, Smoothing Plane, Pore Plane. Gages. Chisels—Firmer Chisel. Trusses. Saw Clamps. The Grindstone. Oilstone. Miter Box. The Work Bench.

Care of Tools—-First Requisites. Saws—How to Set. Saw-set Errors. Saw Setting Block. Filing. The Angle of Filing. Filing Pitch. Saw Clamps. Filing Suggestions. The File. Using the File. The Grindstone. In the Use of Grindstones. Correct Way of Holding Tool in Grinding. Care of Stone. Incorrect Way to Hold Tool. Way to Revolve or Turn Grindstone. The Plane. The Gage. Chisels. General Observations.

On the Holding of Tools. The Saw. How to Start a Saw. Sawing on a Line. The First Stroke. The Starting Cut for Cross-cutting. Forcing a Saw. The Stroke. The Chinese Saw. Things to Avoid. The Plane. Angle for Holding Planes. Errors to be Avoided. The Gage. Holding the Gage. The Draw-knife.

Molding, Re-entering Angle, Rafter, Scarfing, Scotia Molding, Sill, Skewback, Spandrel, Strut, Stud, Stile, Tie Beam, Timber, Trammel, Turret, Transom, Valley Roof.

VIII. Drawing and Its Utility

Fundamentals in Drawing. Representing Objects. Forming Lines and Shadows. Analysis of Lines and Shadings. How to Show Plain Surfaces. Concave Surfaces. Convex Surfaces. Shadows from a Beam. Flat Effects. The Direction of Light. Raised Surfaces. Depressed Surfaces. Full Shading. Illustrating Cube Shading. Shading Effect. Heavy Lines. Perspectives. True Perspective of a Cube. Isometric Cube. Flattened Perspective. Technical Designations. Sector and Segment. Terms of Angles. Circles and Curves. Irregular Curves. Ellipses and Ovals. Focal Points. Produced Line. Spirals, Perpendicular and Vertical. Signs to Indicate Measurement. Definitions. Abscissa. Angle. Apothegm. Apsides or Apsis. Chord. Cycloid. Conoid. Conic Section. Ellipsoid. Epicycloid. Evolute. Flying Buttress. Focus. Gnomes. Hexagon. Hyperbola. Hypothenuse. Incidental. Isosceles. Triangle. Parabola. Parallelogram. Pelecoid. Polygons. Pyramid. Rhomb. Sector. Segment. Sinusoid. Tangent. Tetrahedron. Vertex.

IX. Moldings, with Practical Illustrations in Embellishing Work

Moldings. The Basis of Moldings. The Simplest Moldings.[Pg iv] The Astragal. The Cavetto. The Ovolo. The Torus. The Apothegm. The Cymatium. The Ogee. Ogee Recta. Ogee Reversa. The Reedy. The Casement. The Roman-Doric Column. Lesson from the Doric Column. Applying Molding. Base. Embellishments. Straight-faced Molding. Plain Molding. Base. Diversified Uses. Sha-

87. Inclave

88. Interlacing arch

89. Invected

90. Inverted arch

91. Keystone

92. King post

93. Label

94. Louver

95. Lintel

96. Lug

97. M-roof

98. Mansard roof

99. Newel post

100. Parquetry

101. Peen, or pein

102. Pendant

103. Pentastyle

104. Pedestal

105. Pintle

106. Portico

107. Plate

108. Queen post

109. Quirk molding

110. Re-entering

111. Rafter

112. Scarfing

113. Scotia molding

114. Sill

115. Skew back

116. Spandrel

117. Strut

150. Drawing spiral

151. Abscissa

152. Angle

153. Apothegm

154. Apsides, or apsis

155. Chord

156. Convolute

157. Conic sections

158. Conoid

159. Cycloid

160. Ellipsoid

161. Epicycloid

162. Evolute

163. Focus

164. Gnome

165. Hyperbola

167. Hypothenuse

168. Incidence

169. Isosceles triangle

170. Parabola

171. Parallelogram

172. Pelecoid

173. Polygons

174. Pyramid

175. Quadrant

176. Quadrilaterale

177. Rhomb

178. Sector

179. Segment

180. Sinusoid

181. Tangent

182. Tetrahedron

183. Vertex

184. Volute

185. Band (molding)e

186. Astragal (molding)

187. Cavetto (molding)

188. Ovolo (molding)

189. Torus (molding)

190. Apophyges (molding)

191. Cymatium (molding)

192. Ogee-recta (molding)

193. Ogee-reversa (molding)

194. Bead (molding)

195. Casement (molding)

196. The Doric column

197. Front of cabinet

198. Facia board

199. Molding on facia board

200. Ogee-recta on facia

201. Trim below facia

202. Trim below ogee

203. Trim above base

204. Trim above base molding

205. Shadows cast by plain moldings

206. Mortise and tenon joint

207. Incorrect mortising

208. Steps in mortising

209. The shoulders of tenons

210. Lap-and-butt joint

211. Panel joint

212. Scarfing

CARPENTRY

A PRACTICAL COURSE, WHICH TELLS IN CONCISE AND SIMPLE FORM "HOW TO DO IT"

INTRODUCTORY

Carpentry is the oldest of the arts, and it has been said that the knowledge necessary to make a good carpenter fits one for almost any trade or occupation requiring the use of tools. The hatchet, the saw, and the plane are the three primal implements of the carpenter. The value is in knowing how to use them.

The institution of Manual Training Schools everywhere is but a tardy recognition of the value of systematic training in the use of tools. There is no branch of industry which needs such diversification, in order to become efficient.

The skill of the blacksmith is centered in his ability to forge, to weld, and to temper; that of the machinist depends upon the callipered dimensions of his product; the painter in his taste for harmony; the mason on his ability to cut the stone accurately; and the plasterer to produce a uniform surface. But the carpenter must, in order to be an expert, combine all these qualifications, [Pg 2] in a greater or less degree, and his vocation may justly be called the King of Trades. Rightly, therefore, it should be cultivated in order to learn the essentials of manual training work.

But there is another feature of the utmost importance and value, which is generally overlooked, and on which there is placed too little stress, even in many of the manual training schools. The training of the mind has been systematized so as to bring into operation the energies of all the brain cells. Manual training to be efficient should, at the same time, be directed into such channels as will most widely stimulate the muscular development of the child, while at the same time cultivating his mind.

There is no trade which offers such a useful field as carpentry. It may be said that the various manual operations bring into play every muscle of the body.

The saw, the plane, the hammer, the chisel, each requires its special muscular energy. The carpenter, unlike the blacksmith, does not put all his brawn into his shoulders, nor develop his torso at the expense of his other muscles, like the mason. It may also be said that, unlike most other occupations, the carpenter has both out-of-door and indoor exercise, so that he is at all times able to follow his occupation, summer or [Pg 3] winter, rain or shine; and this also further illustrates the value of this branch of endeavor as a healthful recreation.

It is the aim of this book to teach boys the primary requirements—not to generalize—but to show how to prepare and how to do the work; what tools and materials to use; and in what manner the tools used may be made most serviceable, and used most advantageously.

It would be of no value to describe and illustrate how a bracket is made; or how the framework of a structure is provided with mortises and tenons in order to hold it together. The boy must have something as a base which will enable him to design his own creations, and not be an imitator; his mind must develop with his body. It is the principal aim of this book to give the boy something to think about while he is learning how to bring each individual part to perfection.

If the boy understands that there is a principle underlying each structural device; that there is a reason for making certain things a definite way, he is imbued with an incentive which will sooner or later develop into an initiative of his own.

It is this phase in the artisan's life which determines whether he will be merely a machine or an intelligent organism.

This work puts together in a simple, concise [Pg 4] form, not only the fundamentals which every mechanic should learn to know, but it defines every structural form used in this art, and illustrates all terms it is necessary to use in the employment of carpentry. A full chapter is devoted to drawings practically applied. All terms are diagrammed and defined, so that the mind may readily grasp the ideas involved.

Finally, it will be observed that every illustration has been specially drawn for this book. We have not adopted the plan usually followed in books of this class, of taking stock illustrations of manufacturers' tools and devices, nor have we thought it advisable to take a picture of a tool or a machine and then write a description around it. We have illustrated the book to explain *"how to do the work"*; also, to teach the boy what the trade requires, and to give him the means whereby he may readily find the form of every device, tool, and structure used in the art.

[Pg 5]

CARPENTRY FOR BOYS

CHAPTER I

TOOLS AND THEIR USES

Knowledge of Tools.—A knowledge of tools and their uses is the first and most important requirement. The saw, the plane, the hatchet and the hammer are well known to all boys; but how to use them, and where to use the different varieties of each kind of tool, must be learned, because each tool grew out of some particular requirement in the art. These uses will now be explained.

A Full Kit of Tools.—A kit of tools necessary for doing any plain work should embrace the following:

1. A Hatchet.
2. A Claw Hammer—two sizes preferred.
3. Cross-cut Saw, 20 inches long.
4. Rip Saw, 24 inches long.
5. Wooden Mallet.
6. Jack Plane.
7. Smoothing Plane.
8. Compass Saw.
9. Brace.
10. Bits for Brace, ranging from ¼ inch to 1 inch diameter.

11. Several small Gimlets.
12. Square.
13. Compass.
14. Draw-knife.
15. Rule.
16. Two Gages.
17. Set of Firmer Chisels.
18. Two Mortising Chisels.
19. Small Back Saw.
20. Saw Clamps.
21. Miter Box.
22. Bevel Square.
23. Small Hand Square.
24. Pliers.
25. Pair of Awls.
26. Hand Clamps.
27. Set Files.
28. Glue Pot.
29. Oil Stone.
30. Grindstone.
31. Trusses.
32. Work Bench.
33. Plumb Bob.
34. Spirit Level.

[Pg 6]

The Hatchet.—The hatchet should be ground with a bevel on each side, and not on one side only, as is customary with a plasterer's lathing hatchet, because the blade of the hatchet is used for trimming off the edges of boards. Unless ground off with a bevel on both sides it cannot be controlled to cut accurately. A light hatchet is preferable to a heavy one. It should never be used for nailing purposes, except in emergencies. The pole of the hammer—that part which is generally used to strike the nail with—is required in order to properly balance the hatchet when used for trimming material.

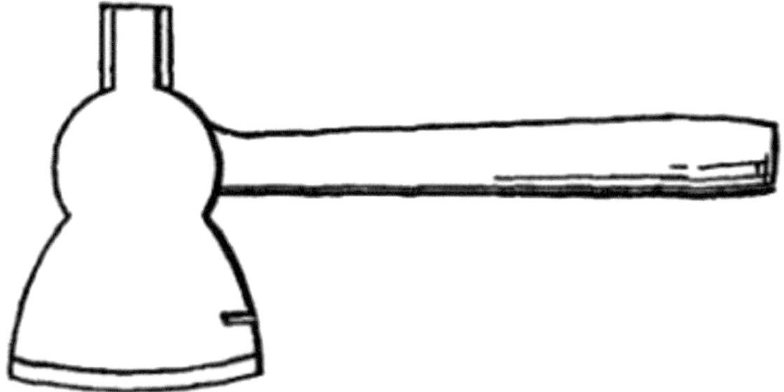

Fig. 2.

The Claw Hammer.—This is the proper tool for driving nails and for drawing them out. Habits should be formed with the beginner, which will be of great service as the education proceeds. [Pg 7] One of these habits is to persist in using the tool for the purpose for which it was made. The expert workman (and he becomes expert because of it) makes the hammer do its proper work; and so with every other tool.

Fig. 3.

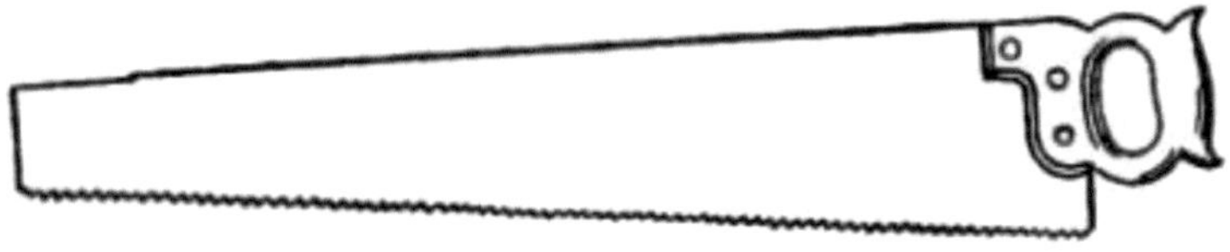

Fig. 4.

About Saws. — There are four well-defined kinds. First, a long, flat saw, for cross-cutting. Second, a slightly larger saw for ripping purposes. Third, a back saw, with a rib on the rear edge to hold the blade rigid, used for making tenons; and, fourth, a compass or key-hole saw.

>[Pg 8]

Cross-cuts. — The difference between a cross-cut and a rip saw is, that in the latter the teeth have less pitch and are usually larger than in the cross-cut saw. The illustrations (Figs. 13 and 14) will distinctly show the difference in the teeth. When a cross-cut saw is used for ripping along the grain of the wood, the teeth, if disposed at an angle, will ride over the grain or fiber of the wood, and refuse to take hold or bite into the wood. On the other hand, if the rip saw is used for cross-cutting purposes, the saw kerf will be rough and jagged.

Fig. 5.

The back saw is used almost exclusively for making tenons, and has uniformly fine teeth so as to give a smooth finish to the wood.

Planes. — The plane may be called the æsthetic tool in the carpenter's kit. It is the most difficult tool to handle and the most satisfactory when thoroughly mastered. How to care for and [Pg 9] handle it will be referred to in a subsequent chapter. We are now concerned with its uses only. Each complete kit must have three distinct planes, namely, the jack plane, which is for taking off the rough saw print surface of the board. The short smoothing plane, which is designed to even up the inequalities made by the jack plane; and the long finishing plane, or fore plane, which is intended to straighten the edges of boards or of finished surfaces.

Fig. 6. Jack plane bit

The Jack Plane.—This plane has the cutting edge of its blade ground so it is slightly curved (Fig. 6), because, as the bit must be driven out so it will take a deep bite into the rough surface of the wood, the curved cutting edge prevents the corner edges of the bit from digging into the planed surface.

On the other hand, the bits of the smoothing and finishing planes are ground straight across their cutting edges. In the foregoing we have not enumerated the different special planes, designed [Pg 10] to make beads, rabbets, tongues and grooves, but each type is fully illustrated, so that an idea may be obtained of their characteristics. (Fig. 6*a*).

Gages.—One of the most valuable tools in the whole set is the gage, but it is, in fact, the least known. This is simply a straight bar, with a sharpened point projecting out on one side near its end, and having an adjustable sliding head or cheekpiece. This tool is indispensable in making mortises or tenons, because the sharpened steel point which projects from the side of the bar, serves to outline and define the edges of the mortises or tenons, so that the cutting line may readily be followed.

Fig. 6a. Fore-plane bit

This is the most difficult tool to hold when in use, but that will be fully explained under its proper head. Each kit should have two, as in making mortises and tenons one gage is required for each side of the mortise or tenon.

Chisels.—Two kinds are found in every kit—one [Pg 11] called the firmer (Fig. 7) and the mortising chisel. The firmer has a flat body or blade, and a full set ranges in width from three-eighths of an inch to two inches. The sizes most desirable and useful are the one-half inch, the inch and the inch-and-a-half widths. These are used for trimming out cross grains or rebates for setting door locks and hinges and for numerous other uses where sharp-end tools are required.

Fig. 7.

The Mortising Chisel.—The mortising chisel (Fig. 7a), on the other hand, is very narrow and thick, with a long taper down to the cutting edge. They are usually in such widths as to make them stock sizes for mortises. Never, under any circumstances, use a hammer or hatchet for driving chisels. The mallet should be used invariably.

Fig. 7a.

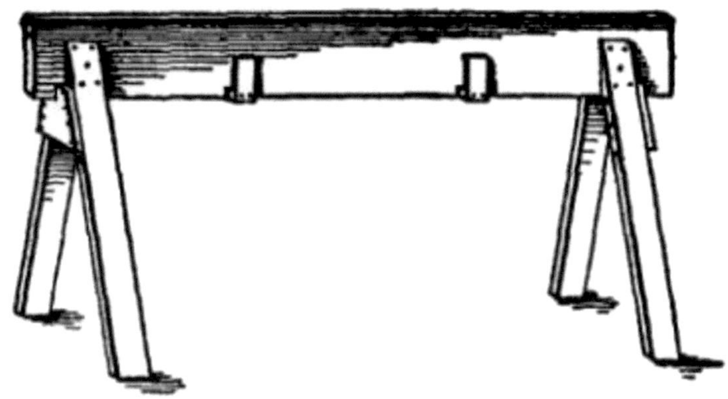

Fig. 8.

Trusses.—There should be at least two, each three feet in length and twenty inches in height.

Saw Clamps.—These are necessary adjuncts, and should be made of hard wood, perfectly [Pg 12] straight and just wide enough to take in the narrow back saw. The illustration shows their shape and form.

The Grindstones.—It is better to get a first-class stone, which may be small and rigged up with a foot treadle. A soft, fine-grained stone is most serviceable, and it should have a water tray, and never be used excepting with plenty of water.

An Oil Stone is as essential as a grindstone. For giving a good edge to tools it is superior to a water stone. It should be provided with a top, and covered when not in use, to keep out dust [Pg 13] and grit. These are the little things that contribute to success and should be carefully observed.

The Miter Box.—This should be 14 inches long and 3" by 3" inside, made of hard wood ¾" thick. The sides should be nailed to the bottom, as shown.

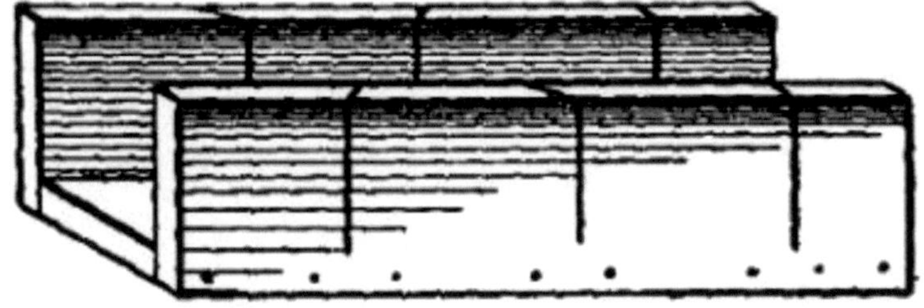

Fig. 9.

The Work Bench. — In its proper place we show in detail the most approved form of work bench, fitted with a tool rack to hold all the tools, conveniently arranged. In this chapter we are more particularly concerned with the uses of tools than their construction; and we impress on boys the necessity of having a place for everything, and that every tool should be kept in its proper place. A carpenter's shop filled with chips, shavings and other refuse is not a desirable place for the indiscriminate placing of tools. If correct habits are formed at the outset, by carefully putting each tool in its place after using, it will save many an hour of useless hunting and annoyance.

One of the most important things in laying off [Pg 14] work, for instance, on trusses, is the disposition of the saw and square. Our illustration shows each truss with side cleats, which will permit the user temporarily to deposit the saw or the square so that it will be handy, and at the same time be out of the way of the work and prevent either of the tools from being thrown to the floor.

In the same way, and for the same purpose, the work bench has temporary holding cleats at the end and a shelf in front, which are particularly desirable, because either a saw or a square is an encumbrance on a work bench while the work is being assembled, and tools of this kind should not be laid flat on a working surface, nor should they be stood in a leaning position against a truss or work bench.

Strictly observe these fundamentals — Never place a tool with the cutting edge toward you. Always have the racks or receptacles so made that the handle may be seized. Don't put a tool with an exposed cutting edge above or below another tool in such a manner that the hand or the tool you are handling can come into contact

34

with the edge. Never keep the nail or screw boxes above the work bench. They should always be kept to one side, to prevent, as much as possible, the bench from becoming a depository for nails. Keep the top of the bench free from tools. Always [Pg 15] keep the planes on a narrow sub-shelf at the rear of the bench.

If order was Heaven's first law, it is a good principle to apply it in a workman's shop, and its observance will form a habit that will soon become a pleasure to follow.

[Pg 16]

CHAPTER II

HOW TO GRIND AND SHARPEN TOOLS

Care of Tools.—Dull tools indicate the character of the workman. In an experience of over forty years, I have never known a good workman to keep poorly sharpened tools. While it is true that the capacity to sharpen tools can be acquired only by practice, correct habits at the start will materially assist. In doing this part of the artisan's work, it should be understood that there is a right as well as a wrong way.

There is a principle involved in the sharpening of every tool, which should be observed. A skilled artisan knows that there is a particular way to grind the bits of each plane; that the manner of setting a saw not only contributes to its usefulness, but will materially add to the life of the saw; that a chisel cannot be made to do good work unless its cutting edge is square and at the right working angle.

First Requisite.—A beginner should never attempt a piece of work until he learns how the different tools should be sharpened, or at least learn the principle involved. Practice will make perfect.

[Pg 17] Saws.—As the saw is such an important part of the kit, I shall devote some space to the subject. *First*, as to setting the saw. The object of this is to make the teeth cut a wider kerf than the thickness of the blade, and thereby cause the saw to travel freely. A great many so-called "saw sets" are found in the market, many of them built on wrong principles, as will be shown, and these are incapable of setting accurately.

Fig. 10. Fig.10a.

How to Set. — To set a saw accurately, that is, to drive out each tooth the same distance, is the first requirement, and the second is to bend out the whole tooth, and not the point only.

In the illustration (Fig. 10), the point is merely bent out. This is wrong. The right way is shown [Pg 18] in Fig. 10a. The whole tooth is bent, showing the correct way of setting. The reasons for avoiding one way and following the other are: First, that if the point projects to one side, each point or tooth will dig into the wood, and produce tooth prints in the wood, which make a roughened surface. Second, that if there are inequalities in setting the teeth (as is sure to be the case when only the points are bent out), the most exposed points will first wear out, and thereby cause saw deterioration. Third, a saw with the points sticking out causes a heavy, dragging cut, and means additional labor. Where the whole body of the tooth is bent, the saw will run smoothly and easily through the kerf and produce a smooth-cut surface.

36

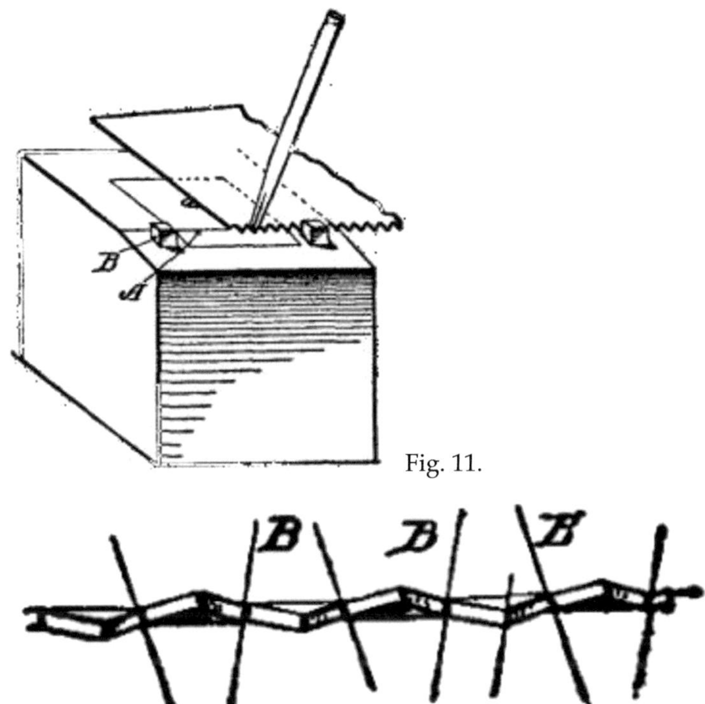

Fig. 11.

Fig. 12.

Our illustration (Fig. 11) shows a very simple setting block, the principal merit of which is that any boy can make it, and in the use of which he cannot go wrong in setting a tooth.

Simple Saw Setter.—Take a block of wood, a 4 by 4 inch studding, four inches long. Get a [Pg 19] piece of metal one-half inch thick and two inches square. Have a blacksmith or machinist bore a quarter-inch hole through it in the center and countersink the upper side so it may be securely fastened in a mortise in the block, with its upper side flush with the upper surface of the block. Now, with a file, finish off one edge, going back for a quarter of an inch, the angle at A to be about 12 degrees.

Fig. 13. Rip-Saw

Filing Angles.—In its proper place will be shown how you may easily calculate and measure degrees in work of this kind. Fig. 12 shows an approximation to the right angle. B, B (Fig. 11) should be a pair of wooden pegs, driven into the wooden block on each side of the metal piece. The teeth of the saw rest against the pegs so that they serve as a guide or a gage, and the teeth of the saw, therefore, project over the inclined part (B) of the metal block. Now, with [Pg 20] an ordinary punch and a hammer, each alternate tooth may be driven down until it rests flat on the inclined face (A), so that it is impossible to set the teeth wrongly. When you glance down the end of a properly set saw, you will see a V-shaped channel, and if you will place a needle in the groove and hold the saw at an angle, the needle will travel down without falling out.

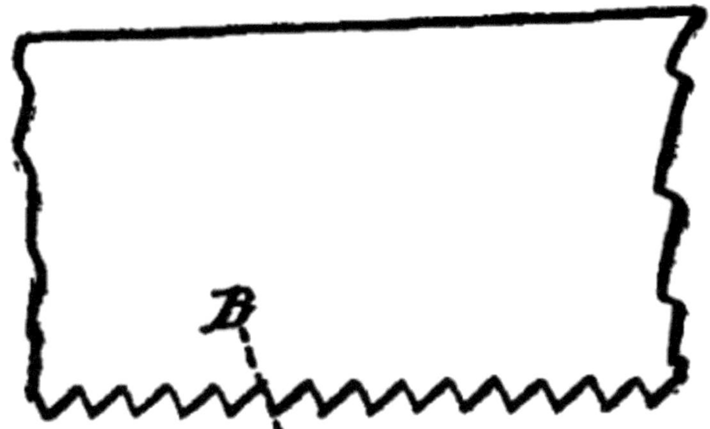

Fig. 14. cross-cut

Filing.—The next step is the filing. Two things must be observed: the pitch and the angle. By pitch is meant the inclination of the teeth. Note the illustration (Fig. 13), which shows the teeth of a rip saw. You will see at A that the pitch of the tooth is at right angles to the edge of the saw. In Fig. 14, which shows the teeth of a cross-cut saw, the pitch (B) is about 10 degrees off. The teeth of the rip saw are also larger than those of the cross-cut.

The Angle of Filing.—By angle is meant the cutting position of the file. In Fig. 12, the lines [Pg 21] B represent the file disposed at an angle of 12 degrees, not more, for a rip saw. For a cross-cut the angle of the file may be less.

Saw Clamps.—You may easily make a pair of saw clamps as follows:

Take two pieces of hard wood, each three inches wide, seven-eighths of an inch thick, and equal in length to the longest saw. Bevel one edge of each as shown in A (Fig. 15), so as to leave an edge (B) about one-eighth of an inch thick. At one end cut away the corner on the side opposite the bevel, as shown at C, so the clamps will fit on the saw around the saw handle.

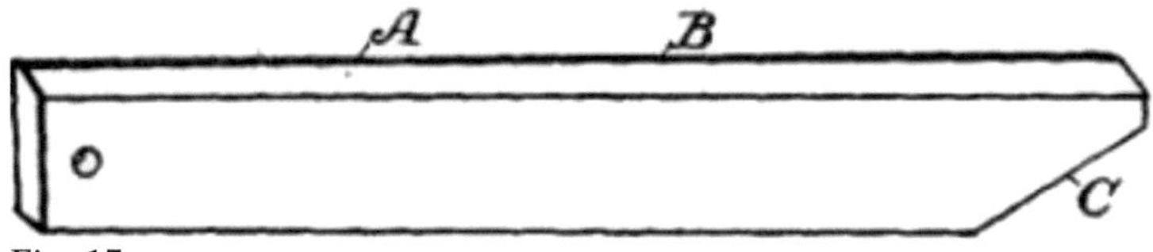

Fig. 15.

When the saw is placed between these clamps and held together by the jaws of the vise, you are ready for the filing operation. Observe the following *filing suggestions*: Always hold the file horizontal or level. In filing, use the whole length of the file. Do the work by a slow, firm sweep.

Do not file all of the teeth along the saw at one operation, but only the alternate teeth, so as to [Pg 22] keep the file at the same angle, and thus insure accuracy; then turn the saw and keep the file constantly at one angle for the alternate set of teeth.

Give the same number of strokes, and exert the same pressure on the file for each tooth, to insure uniformity. Learn also to make a free, easy and straight movement back and forth with the file.

The File. — In order to experiment with the filing motion, take two blocks of wood, and try surfacing them off with a file. When you place the two filed surfaces together after the first trial both will be convex, because the hands, in filing, unless you exert the utmost vigilance, will assume a crank-like movement. The filing test is so to file the two blocks that they will fit tightly together without rolling on each other. Before shaping and planing machines were invented, machinists were compelled to plane down and accurately finish off surfaces with a file.

In using the files on saws, however small the file may be, one hand should hold the handle and the other hand the tip of the file.

A file brush should always be kept on hand, as it pays to preserve files by cleaning them.

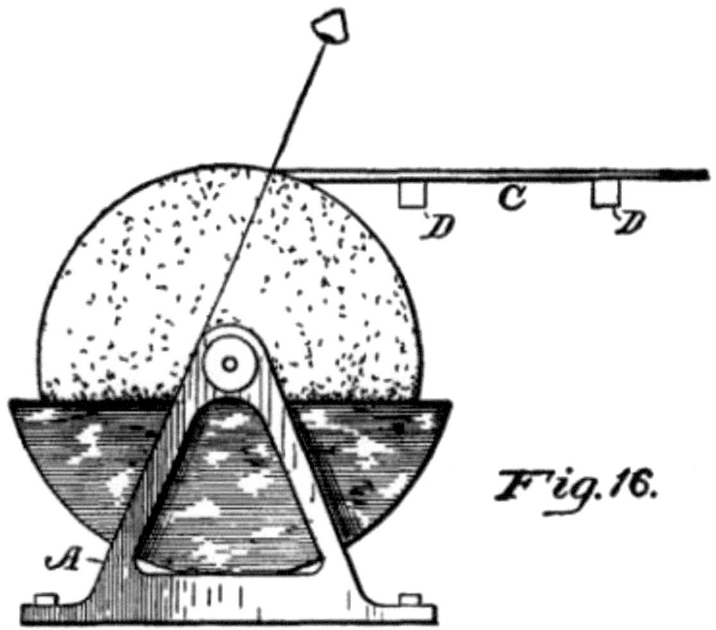

Fig. 16.

The Grindstone. — As most of the tools require a grindstone for sharpening purposes, an illustration is given as a guide, with a diagram to show the proper grinding angle. In Fig. 16 the upright [Pg 23] (A) of the frame serves as a line for the eye, so that if the point of the tool is brought to the sight line, and the tool (C) held level, you will always be able to maintain the correct angle. There is no objection to providing a rest, for instance, like the cross bars (D, D), but the artisan disdains such contrivances, and he usually avoids them for two reasons: First, because habit enables him to hold the tool horizontally; and, second, by holding the tool firmly in the hand he has better control of it. There is only one thing which can be said in favor of a rest, and [Pg 24] that is, the stone may be kept truer circumferentially, as all stones have soft spots or sides.

In the Use of Grindstones. — There are certain things to avoid and to observe in the use of stones. Never use one spot on the stone, however narrow the tool may be. Always move the tool from side to side. Never grind a set of narrow tools successively. If you have chisels to grind intersperse their grinding with plane bits, hatchet or other broad cutting tools, so as to prevent the stone from having grooves therein. Never use a tool on a stone unless you have water in the tray.

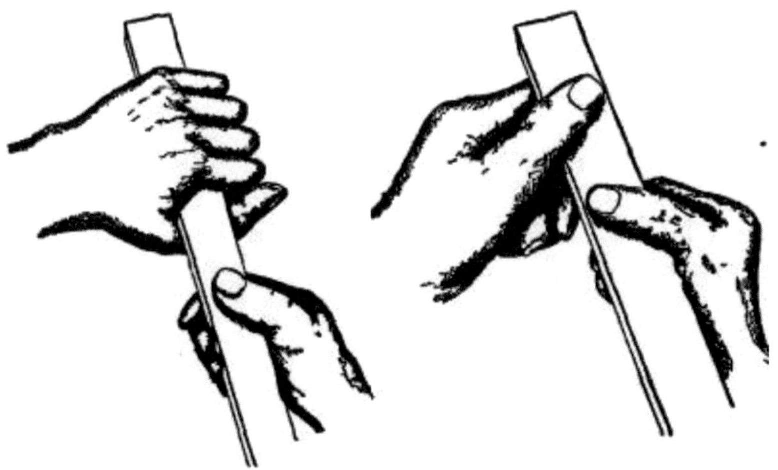

Fig. 17. Correct.   Fig. 18. Incorrect.

Correct Way to Hold Tool for Grinding.—There is a correct way to hold each tool; see illustration (Fig. 17). The left hand should grasp the tool firmly, near the sharp edge, as shown, and the right hand should loosely hold the tool behind [Pg 25] the left hand. There is a reason for this which will be apparent after you grind a few tools. The firm grasp of the left hand gives you absolute control of the blade, so it cannot turn, and when inequalities appear in the grindstone, the rigid hold will prevent the blade from turning, and thus enable you to correct the inequalities of the stone. Bear in mind, the stone should be taken care of just as much as the tools. An experienced workman is known by the condition of his tools, and the grindstone is the best friend he has among his tools.

Incorrect Way to Hold Tool for Grinding.—The incorrect way of holding a tool is shown in Fig. 18. This, I presume, is the universal way in which the novice takes the tool. It is wrong for the reason that the thumbs of both hands are on top of the blade, and they serve as pivots on which the tool may turn. The result is that the corners of the tool will dig into the stone to a greater or less degree, particularly if it has a narrow blade, like a chisel.

Try the experiment of grinding a quarter-inch chisel by holding it the incorrect way; and then grasp it firmly with the left hand, and you will at once see the difference.

The left hand serves both as a vise and as a [Pg 26] fulcrum, whereas the right hand controls the angle of the tool.

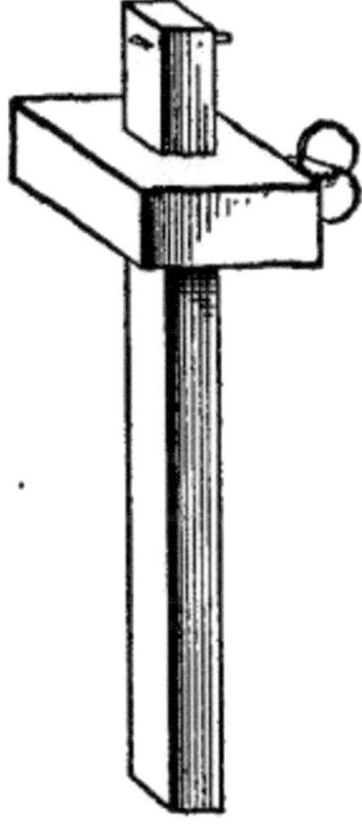

Fig. 19.

These remarks apply to all chisels, plane bits and tools of that character, but it is obvious that a drawknife, which is always held by the handles in grinding, and hatchets, axes and the like, cannot be held in the same manner.

A too common error is to press the tool too hard on the stone. This is wrong. Do not try to force the grinding.

Then, again, it is the practice of some to turn the stone away from the tool. The stone should always move toward the tool, so as to prevent forming a feather edge.

[Pg 27] The Plane. — Indiscriminate use of planes should be avoided. Never use the fore or smoothing planes on rough surfaces. The jack plane is the proper tool for this work. On the other hand, the fore plane should invariably be used for straightening the edges of boards, or for fine surfacing purposes. As the jack plane has its bit ground with a curved edge, it is admirably adapted for taking off the rough saw print surface.

The Gage. — The illustration (Fig. 19) shows one of the most useful tools in the kit. It is used to scribe the thickness of the material which is to be dressed down, or for imprinting the edges of tenons and mortises. Two should be provided in every kit, for convenience.

The scribing point should be sharpened with a file, the point being filed to form a blade, which is at right angles to the bar, or parallel with the movable cheekpiece.

Chisels. — I have already pointed out, in general, how to hold tools for grinding purposes, this description applying particularly to chisels, but several additional things may be added.

Always be careful to grind the chisel so its cutting edge is square with the side edge. This will be difficult at first, but you will see the value of this as you use the tool. For instance, in making [Pg 28] rebates for hinges, or recesses and mortises for locks, the tool will invariably run crooked, unless it is ground square.

The chisel should never be struck with a hammer or metal instrument, as the metal pole or peon of the hammer will sliver the handle. The wooden mallet should invariably be used.

General Observations. — If the workman will carefully observe the foregoing requirements he will have taken the most important steps in the knowledge of the art. If he permits himself to commence work without having his tools in first-class condition, he is trying to do work under circumstances where even a skilled workman is liable to fail.

Avoid making for yourself a lot of unnecessary work. The best artisans are those who try to find out and know which is the best tool, or how to make a tool for each requirement, but that tool, to be serviceable, must be properly made, and that means it must be rightly sharpened.

CHAPTER III

HOW TO HOLD AND HANDLE TOOLS

Observation may form part of each boy's lesson, but when it comes to the handling of tools, practice becomes the only available means of making a workman. Fifty years of observation would never make an observer an archer or a marksman, nor would it enable him to shoe a horse or to build a table.

It sometimes happens that an apprentice will, with little observation, seize a saw in the proper way, or hold a plane in the correct manner, and, in time, the watchful boy will acquire fairly correct habits. But why put in useless time and labor in order to gain that which a few well-directed hints and examples will convey?

Tools are made and are used as short cuts toward a desired end. Before the saw was invented the knife was used laboriously to sever and shape the materials. Before planes were invented a broad, flat sharpened blade was used to smooth off surfaces. Holes were dug out by means of small chisels requiring infinite patience and time. Each succeeding tool proclaimed a shorter and an easier way to do a certain thing. [Pg 30] The man or boy who can make a new labor-saving tool is worthy of as much praise as the man who makes two blades of grass grow where one grew before.

Let us now thoroughly understand how to hold and use each tool. That is half the value of the tool itself.

The Saw.—With such a commonplace article as the saw, it might be assumed that the ordinary apprentice would look upon instruction with a smile of derision.

How to Start a Saw.—If the untried apprentice has such an opinion set him to work at the task of cutting off a board accurately on a line. He will generally make a failure of the attempt to start the saw true to the line, to say nothing of following the line so the kerf is true and square with the board.

How to Start on a Line.—The first mistake he makes is to saw *on the line*. This should never be done. The work should be so laid out that the saw kerf is on the discarded side of the material. The saw should cut alongside the line, and *the line should not* be obliterated in the cutting. Material must be left for trimming and finishing.

The First Stroke.—Now, to hold the saw in starting is the difficult task to the beginner. Once mastered it is simple and easy. The only time in [Pg 31] which the saw should be firmly held by the hand is during the initial cut or two; afterwards always hold the handle loosely. There is nothing so tiring as a tightly grasped saw. The saw has but one handle, hence it is designed to be used with one hand. Sometimes, with long and tiresome jobs, in ripping, two hands may be used, but one hand can always control a saw better than two hands.

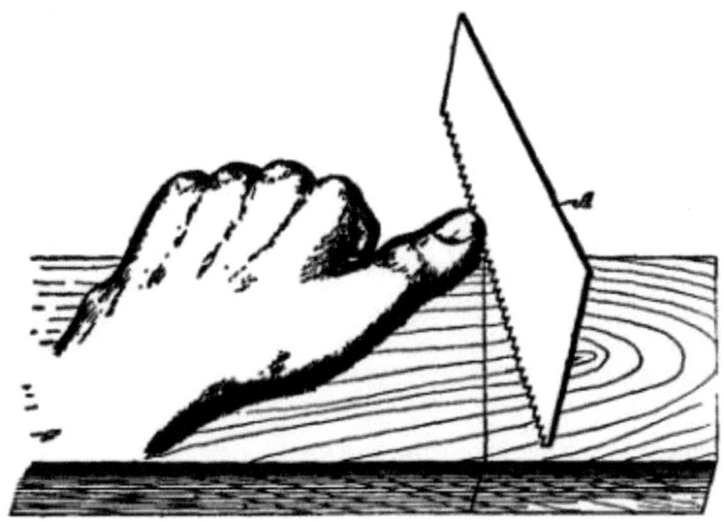

Fig. 20.

The Starting Cut.—In order to make our understanding of the starting cut more explicit, we refer to Fig. 20, in which the thumb of the left hand is shown in the position of a guide—the end of the thumb being held up a sufficient distance to [Pg 32] clear the teeth. In this position you need not fear that the teeth of the saw (A) will ride up over the thumb if you have a firm grasp of the saw handle.

The first stroke should be upwardly, not downwardly. While in the act of drawing up the saw you can judge whether the saw blade is held by the thumb gage in the proper position to cut along the mark, and when the saw moves downwardly for the first cut, you may be assured that the cut is accurate, or at the right place, and the thumb should be kept in its position until two or three cuts are made, and the work is then fairly started.

Fig. 21. Wrong sawing angle.

For Cross-cutting.—For ordinary cross-cutting the angle of the saw should be at 45 degrees. For ripping, the best results are found at less than 45 degrees, but you should avoid flattening down the angle. An incorrect as well as a correct angle are shown in Figs. 21 and 22.

Forcing a Saw.—Forcing a saw through the wood means a crooked kerf. The more nearly the saw is held at right angles to a board, the greater [Pg 33] is the force which must be applied to it by the hand to cause it to bite into the wood; and, on the other hand, if the saw is laid down too far, as shown in the incorrect way, it is a very difficult matter to follow the working line. Furthermore, it is a hard matter to control the saw so that it will cut squarely along the board, particularly when ripping. The eye must be the only guide in the disposition of the saw. Some boys make the saw run in one direction, and others cause it to lean the opposite way. After you have had some experience and know which way you lean, correct your habits by disposing the saw in the opposite direction.

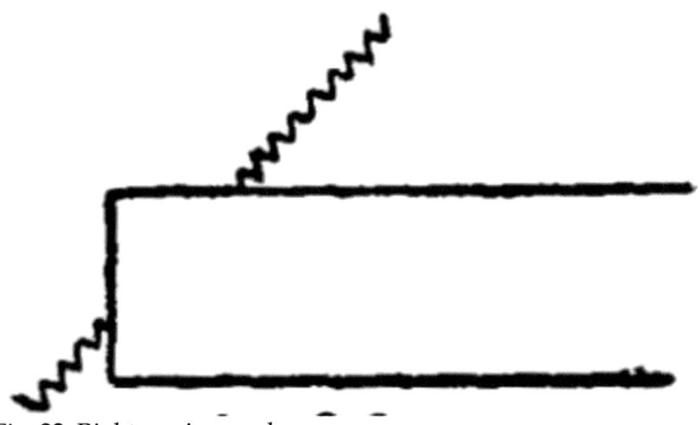

Fig. 22. Right sawing angle.

The Stroke.—Make a long stroke, using the full blade of the saw. Don't acquire the "jerky" style of sawing. If the handle is held loosely, and the saw is at the proper angle, the weight of the saw, together with the placement of the handle on the saw blade, will be found sufficient to make the requisite cut at each stroke.

[Pg 34] You will notice that the handle of every saw is mounted nearest the back edge. (See Fig. 23.) The reason for so mounting it is, that as the cutting stroke is downward, the line of thrust is above the tooth line, and as this line is at an angle to the line of thrust, the tendency is to cause the saw teeth to dig into the wood.

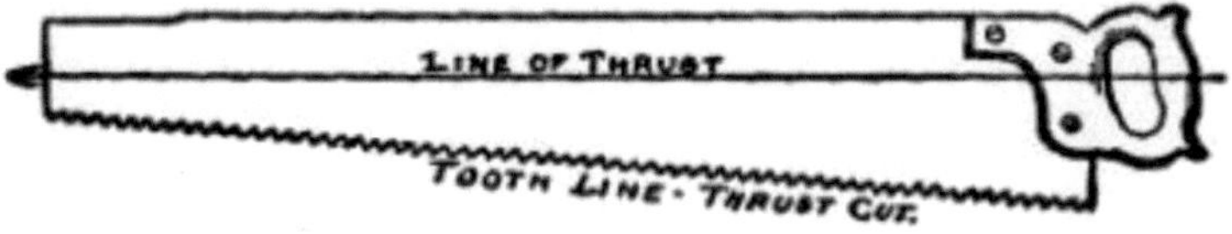

Fig. 23.

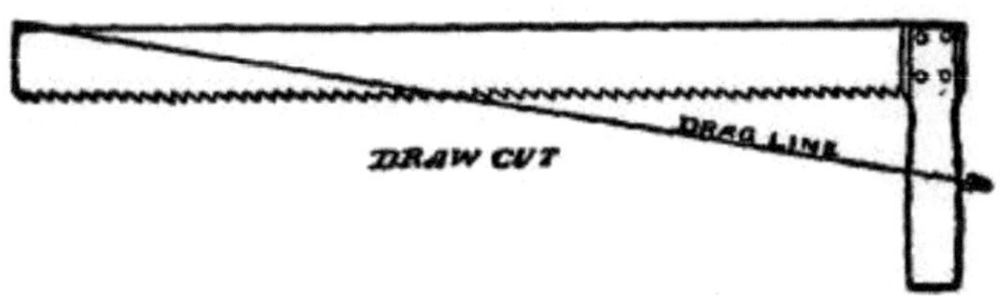

CHINESE SAW. *Fig. 24.*

The Chinese Saw.—This saw is designed to saw with an upward cut, and the illustration (Fig. 24) shows the handle jutting out below the tooth line, in order to cause the teeth to dig into the material as the handle is drawn upwardly. Reference is made to these features to impress upon beginners the value of observation, and to demonstrate the reason for making each tool a particular way.

[Pg 35] Things to Avoid.—Do not oscillate the saw as you draw it back and forth. This is unnecessary work, and shows impatience in the use of the tool. There is such an infinite variety of use for the different tools that there is no necessity for rendering the work of any particular tool, or tools, burdensome. Each in its proper place, handled intelligently, will become a pleasure, as well as a source of profit.

Fig. 25.

The Plane.—The jack plane and the fore plane are handled with both hands, and the smoothing plane with one hand, but only when used for dressing the ends of boards. For other uses both hands are required.

Angles for Holding Planes.—Before commencing to plane a board, always observe the direction in which the grain of the wood runs. This precaution will save many a piece of material, because if the jack plane is set deep it will run into the wood and cause a rough surface, which can [Pg 36] be cured only by an extra amount of labor in planing down.

Never move the jack plane or the smoothing plane over the work so that the body of the tool is in a direct line with the movement of the plane. It should be held at an angle of about 12 or 15 degrees (see Fig. 25). The fore plane should always be held straight with the movement of the plane, because the length of the fore plane body is used as a straightener for the surface to be finished.

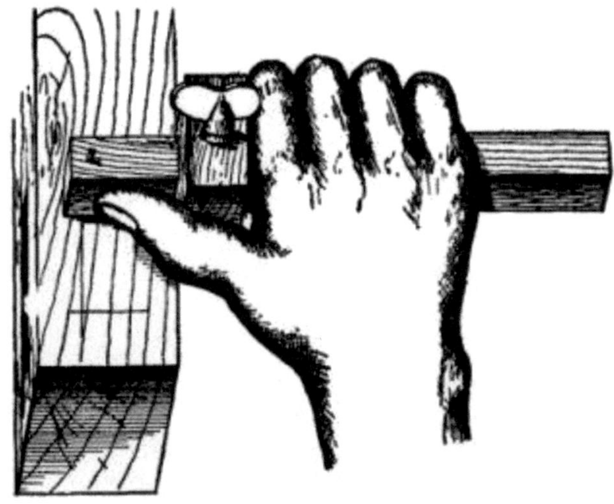

Fig. 26.

Errors to Be Avoided.—Never draw back the plane with the bit resting on the board. This [Pg 37] simply wears out the tool, and if there should be any grit on the board it will be sure to ruin the bit. This applies particularly to the jack plane, but is bad practice with the others as well.

A work bench is a receptacle for all kinds of dirt. Provide a special ledge or shelf for the planes, and be sure to put each plane there immediately after using.

The Gage.—A man, who professed to be a carpenter, once told me that he never used a gage because he could not make it run straight. A few moments' practice convinced him that he never knew how to hold it. The illustration shows how properly to hold it, and the reason why it should so be held follows.

You will observe (Fig. 26) that the hand grasps the stem of the gage behind the cheekpiece, so that the thumb is free to press against the side of the stem to the front of the cheekpiece.

50

Holding the Gage.—The hand serves to keep the cheekpiece against the board, while the thumb pushes the gage forward. The hand must not, under any circumstances, be used to move the gage along. In fact, it is not necessary for the fingers to be clasped around the gage stem, if the forefinger presses tightly against the cheek-piece, since the thumb performs all the operation of moving it along. Naturally, the hand grasps the tool in [Pg 38] order to hold it down against the material, and to bring it back for a new cut.

The Draw-knife.—It is difficult for the apprentice to become ac-customed to handle this useful tool. It is much more serviceable than a hatchet for trimming and paring work. In applying it to the wood always have the tool at an angle with the board, so as to make a slicing cut. This is specially desirable in working close to a line, otherwise there is a liability of cutting over it.

This knife requires a firm grasp—firmness of hold is more im-portant than strength in using. The flat side is used wholly for straight edges, and the beveled side for concave surfaces. It is the intermediate tool between the hatchet and the plane, as it has the characteristics of both those tools. It is an ugly, dangerous tool, more to be feared when lying around than when in use. Put it reli-giously on a rack which protects the entire cutting edge. *Keep it off the bench.*

[Pg 39]

CHAPTER IV

HOW TO DESIGN ARTICLES

Fundamentals of Designing.—A great deal of the pleasure in making articles consists in creative work. This means, not that you shall design some entirely new article, but that its general form, or arrangement of parts, shall have some new or striking feature.

A new design in any art does not require a change in all its parts. It is sufficient that there shall be an improvement, either in some particular point, as a matter of utility, or some change in an artistic direction. A manufacturer in putting out a new chair, or a plow, or an automobile, adds some striking characteristic. This becomes his talking point in selling the article.

The Commercial Instinct.—It is not enough that the boy should learn to make things correctly, and as a matter of pastime and pleasure. The commercial instinct is, after all, the great incentive, and should be given due consideration.

It would be impossible, in a book of this kind, to do more than to give the fundamental principles necessary in designing, and to direct the mind [Pg 40] solely to essentials, leaving the individual to build tip for himself.

First Requirements for Designing.—First, then, let us see what is necessary to do when you intend to set about making an article. Suppose we fix our minds upon a table as the article selected. Three things are necessary to know: First, the use to which it is to be put; second, the dimensions; and, third, the material required.

Assuming it to be the ordinary table, and the dimensions fixed, we may conclude to use soft pine, birch or poplar, because of ease in working. There are no regulation dimensions for tables, except as to height, which is generally uniform, and usually 30 inches. As to the length and width, you will be governed by the place where it is to be used.

If the table top is to have dimensions, say, of 36" × 48", you may lay out the framework six inches less each way, thus giving you a top overhang of three inches, which is the usual practice.

Conventional Styles.—Now, if you wish to depart from the conventional style of making a table you may make variations in the design. For instance, the Chippendale style means slender legs and thin top. It involves some fanciful designs in the curved outlines of the top, and in the crook [Pg 41] of the legs. Or if, on the other hand, the Mission type is preferred, the overhang of the top is very narrow; the legs are straight and heavy, and of even size from top to bottom; and the table top is thick and nearly as broad as it is long. Such furniture has the appearance of massiveness; it is easily made and most serviceable.

Mission Style.—The Mission style of architecture also lends itself to the making of chairs and other articles of furniture. A chair is, probably, the most difficult piece of household furniture to make, because strength is required. In this type soft wood may be used, as the large legs and back pieces are easily provided with mortises and tenons, affording great rigidity when completed. In designing, therefore, you may see how the material itself becomes an important factor.

Cabinets.—In the making of cabinets, sideboards, dressers and like articles, the ingenious boy will find a wonderful field for designing ability, because in these articles fancy alone dictates the sizes and the dimensions of the parts. Not so with chairs and tables. The imagination plays an important part even in the making of drawers, to say nothing of placing them with an eye to convenience and artistic effect.

Harmony of Parts.—But one thing should be observed in the making of furniture, namely, harmony [Pg 42] between the parts. For instance, a table with thin legs and a thick top gives the appearance of a top-heavy structure; or the wrong use of two different styles is bad from an artistic standpoint; moreover, it is the height of refined education if, in the use of contrasting woods, they are properly blended to form a harmonious whole.

Harmonizing Wood.—Imagine a chiffonier with the base of dark wood, like walnut, and the top of pine or maple, or a like light-colored wood. On the other hand, both walnut and maple, for instance, may be used in the same article, if they are interspersed throughout the entire article. The body may be made of dark wood and trimmed throughout with a light wood to produce a fine effect.

[Pg 43]

CHAPTER V

HOW WORK IS LAID OUT

Concrete Examples of Work.—A concrete example of doing any work is more valuable than an abstract statement. For this purpose I shall direct the building of a common table with a drawer in it and show how the work is done in detail.

For convenience let us adopt the Mission style, with a top 36" × 42" and the height 30". The legs should be 2" × 2" and the top 1", dressed. The material should be of hard wood with natural finish, or, what is better still, a soft wood, like birch, which may be stained a dark brown, as the Mission style is more effective in dark than in light woods.

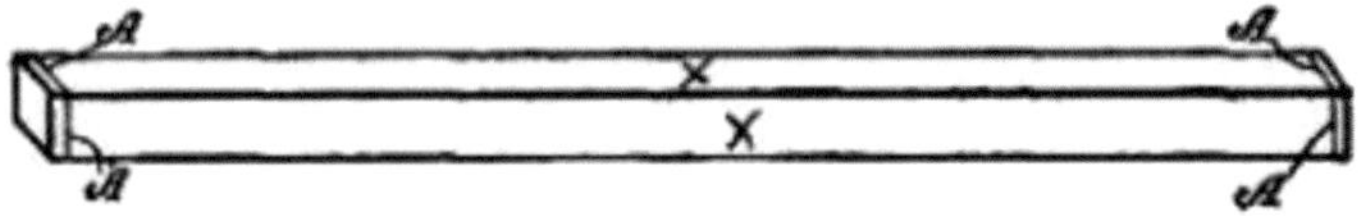

Fig. 27.

Framework.—As we now know the sizes, the first thing is to build the framework. The legs should be dressed square and smoothed down with the fore plane to make them perfectly straight. Now, lay out two mortises at the upper end of each [Pg 44] leg. Follow the illustrations to see how this is done.

Laying Out the Legs.—Fig. 27 shows a leg with square cross marks (A) at each end. These marks indicate the finished length of the leg. You will also see crosses on two sides. These indicate what is called the "work sides." The work sides are selected because they are the finest surfaces on the leg.

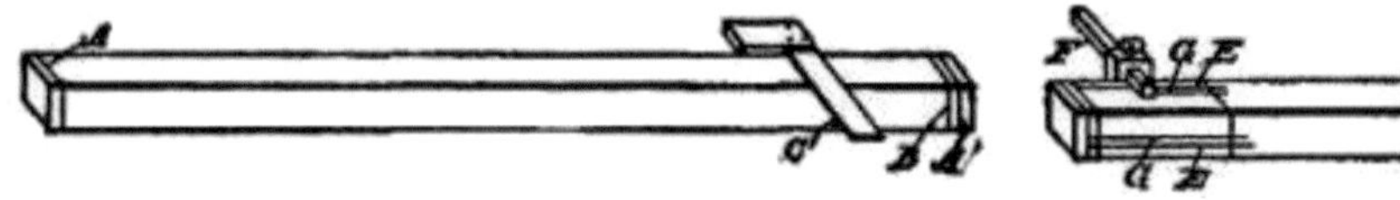

Fig. 28.

The Length of the Mortises.—Then take a small try square (Fig. 28) and add two cross lines (B, C) on each of the inner surfaces, the second line (B) one-half inch from the finish line (A), and the other line (C) seven inches down from the line (A). The side facing boards, hereafter described, are seven inches wide.

When this has been done for all the legs, prepare your gage (Fig. 29) to make the mortise scribe, and, for convenience in illustrating, the leg [Pg 45] is reversed. If the facing boards are 1" thick, and the tenons are intended to be ½" thick, the first scribe line (E) should be

½" from the work side, because the shoulder on the facing board projects out ¼", and the outer surface of the facing board should not be flush with the outer surface of the leg. The second gage line (F) should be 1" from the work side.

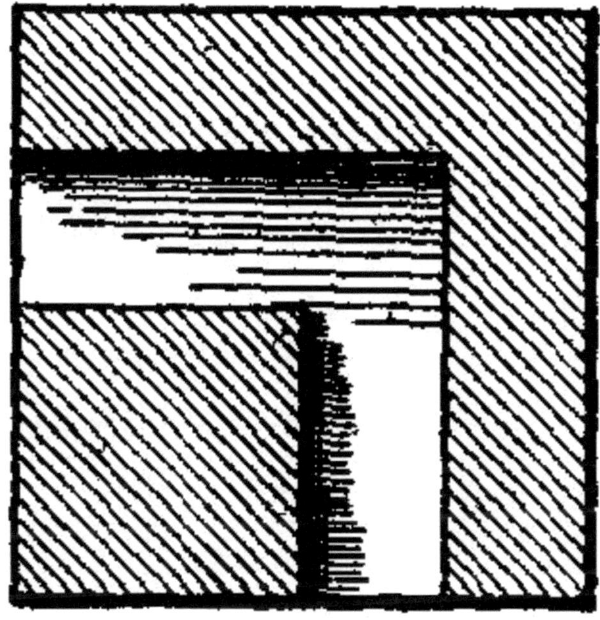

Fig. 30.

The Mortises.—When the mortises have been made they will appear as shown in the enlarged cross section of the leg (Fig. 30), the total depth of each mortise being 1½". The depth of this mortise determines for us the length of the tenons on the facing boards.

The Facing Boards.—These boards are each 1 inch thick and 7 inches wide. As the top of the table is 42 inches long, and we must provide an overhang, say of 2 inches, we will first take off 4 inches for the overhang and 4 inches for the [Pg 46] legs, so that the length of two of the facing boards, from shoulder to shoulder, must be 34 inches; and the other two facing boards 28 inches. Then, as we must add 1½ inches for each tenon, two of the boards will be 37 inches long and two of them 31 inches long.

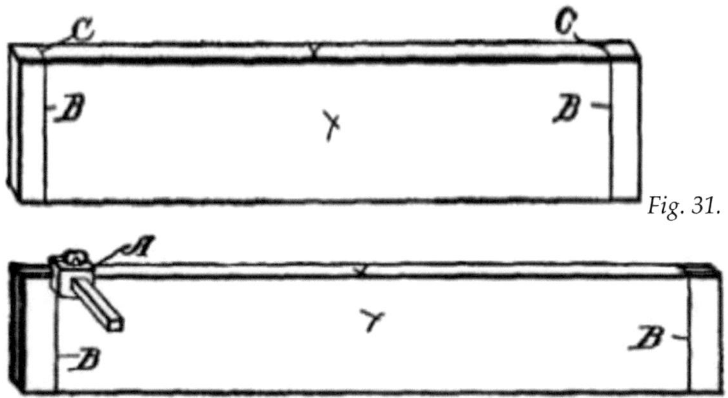

Fig. 31.

Fig. 32.

The illustration (Fig. 31) shows a board marked with the cross lines (B) at each end for the end of the tenons, or the extreme ends of the boards.

The Tenons. — Do not neglect first to select the work side and the working edge of the board. The outer surface and the upper edges are the sides to work from. The cheekpiece (A) of the gage must always rest against the working side.

[Pg 47] The cross marks (B, C) should be made with the point of a sharp knife, and before the small back saw is used on the cross-cuts the lines (B), which indicate the shoulders, should be scored with a sharp knife, as shown in Fig. 33. This furnishes a guide for the saw, and makes a neat finish for the shoulder.

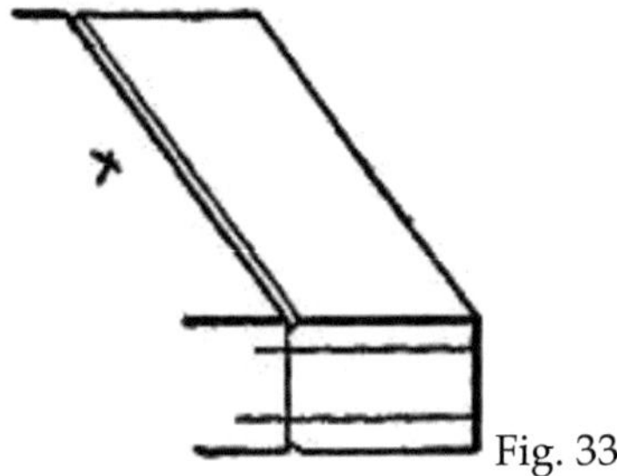

Fig. 33.

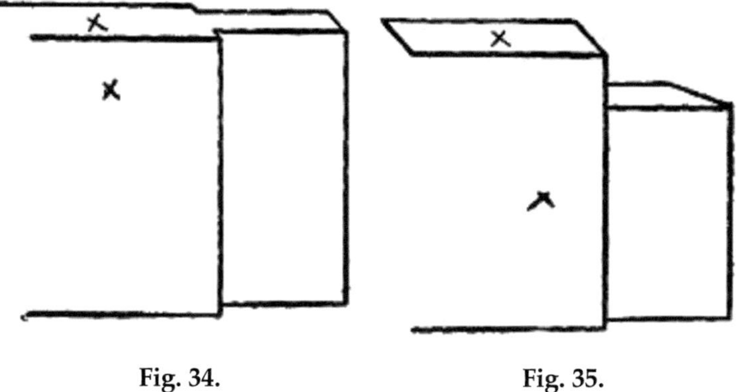

Fig. 34.   Fig. 35.

Tools Used.—The back saw is used for cutting the tenon, and the end of the board appears as [Pg 48] shown in the enlarged Fig. 34. Two things are now necessary to complete the tenons. On the upper or work edge of each board use the gage to mark off a half-inch slice, and then cut away the flat side of the tenon at the end, on its inner surface, so it will appear as shown in Fig. 35.

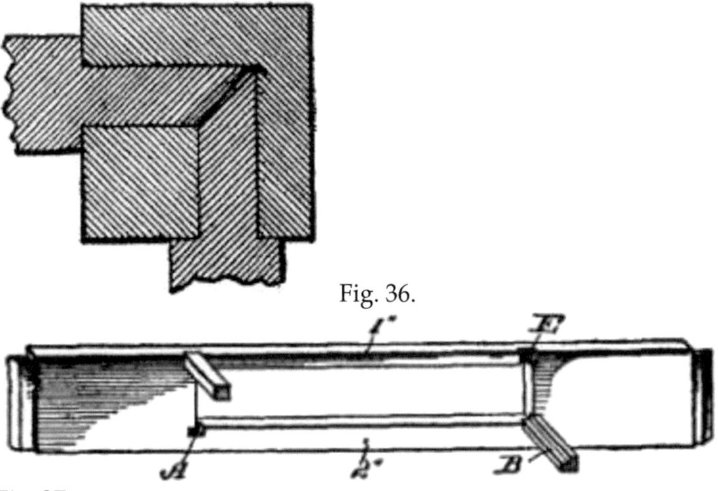

Fig. 36.

Fig. 37.

Chamfered Tenons.—The object of these chamfered or beveled tenons is to permit the ends to approach each other closely within the mortise, as shown in the assembled parts (Fig. 36).

The Frame Assembled.—The frame is now ready to assemble, but before doing so a drawer opening and supports should be made. The ends [Pg 49] of the supports may be mortised into the side pieces or secured by means of gains.

Mortises and tenons are better.

The Drawer Supports.—Take one of the side-facing boards (Fig. 37) and cut a rectangular opening in it. This opening should be 4 inches wide and 18 inches long, so placed that there is 1 inch of stock at the upper margin and 2 inches of stock at the lower margin of the board. At each lower corner make a mortise (A), so that one side of the mortise is on a line with the margin of the opening, and so that it extends a half inch past the vertical margin of the opening.

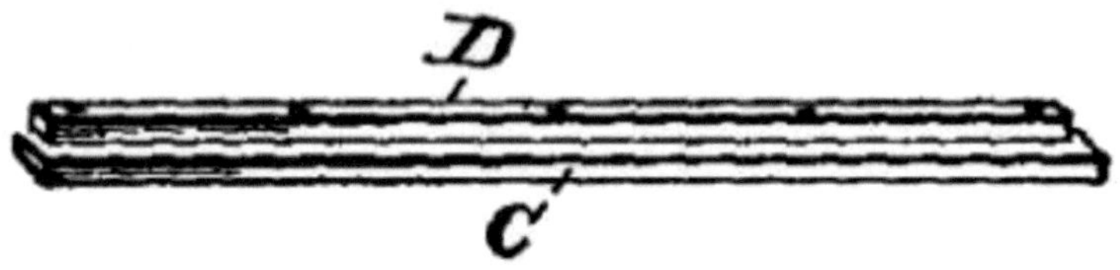

Fig. 38.

You can easily cut a gain (B) in a strip, or, as in Fig. 38, you may use two strips, one (C) an inch wide and a half inch thick, and on this nail a strip (D) along one margin. This forms the guide and rest for the drawer.

At the upper margin of the opening is a rebate or gain (E) at each corner, extending down to the top line of the drawer opening, into which are fitted the ends of the upper cross guides.

[Pg 50] The Table Frame.—When the entire table frame is assembled it will have the appearance shown in Fig. 39, and it is now ready for the top.

The Top.—The top should be made of three boards, either tongued and grooved, or doweled and glued together. In order to give a massive appearance, and also to prevent the end grain of the boards from being exposed, beveled strips may be used to encase the edges. These marginal cleats are ¾ inch thick and 2 inches wide, and joined by beveled ends at the corners, as shown in Fig. 40.

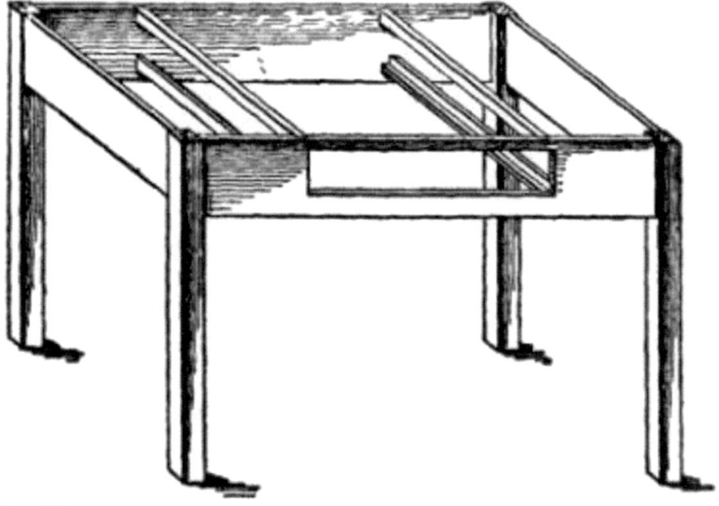

Fig. 39.

The Drawer.—The drawer (Fig. 41) shown in cross section, has its front (A) provided with an overlapping flange (B).

[Pg 51] It is not our object in this chapter to show how each particular article is made, but simply to point out the underlying principles, and to illustrate how the fastening elements, the tenons and mortises, are formed, so that the boy will know the proper steps in their natural order.

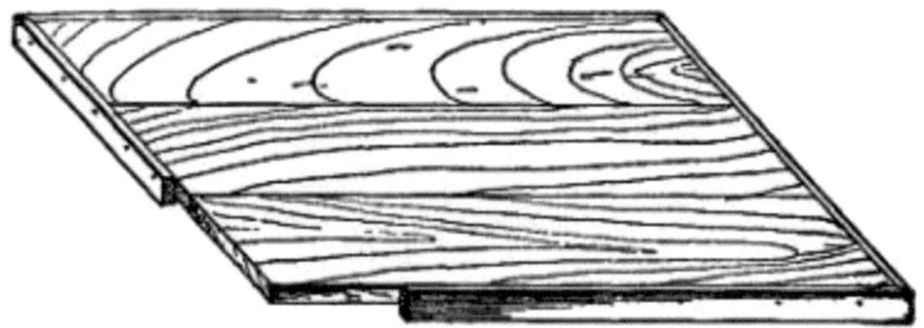

Fig. 40.

How Any Structure Is Built Up.—It should be observed that each structure, however small, is usually built from the base up. Just the same as the more pretentious buildings are erected: First, the sill, then the floor supports, then the posts and top plates, with their connecting girders, and, finally, the roof.

The chapter on House Building will give more detailed illustrations of large structures, and how they are framed and braced. At this point we are more concerned in knowing how to proceed in order to lay out the simple structural details, and if one subject of this kind is fully mastered the complicated character of the article will not be difficult to master. [Pg 52]

Observations About a Box.—As simple a little article as a box frequently becomes a burden to a beginner. Try it. Simply keep in mind one thing; each box has six sides. Now, suppose you want a box with six equal sides—that is, a cubical form—it is necessary to make only three pairs of sides; two for the ends, two for the sides and two for the top and bottom. Each set has dimensions different from the other sets. Both pieces of the set, representing the ends, are square; the side pieces are of the same width as the end pieces, and slightly longer; and the top and bottom are longer and wider than the end pieces.

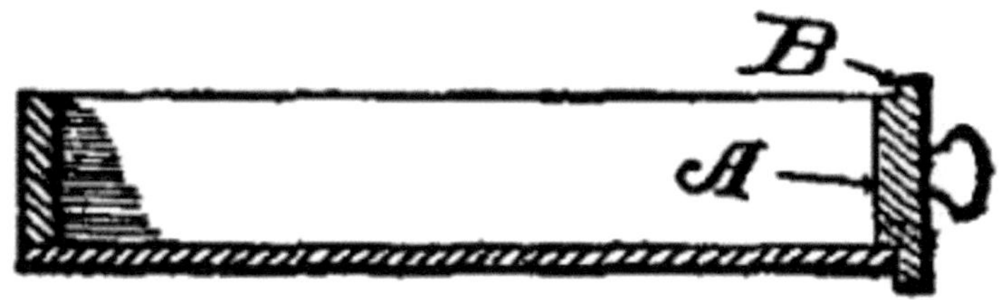

Fig. 41.

A box equal in all its dimensions may be made out of six boards, properly cut. Make an attempt in order to see if you can get the right dimensions.

Joints.—For joining together boards at right angles to each other, such as box corners, drawers and like articles, tenons and mortises should never be resorted to. In order to make fine work the joints should be made by means of dovetails, rabbets [Pg 53] or rebates, or by beveling or mitering the ends.

Beveling and Mitering.—There is a difference in the terms "beveling" and "mitering," as used in the art. In Fig. 42 the joint A is *beveled*, and in Fig. 43 the joint B is *mitered*, the difference being that a bevel is applied to an angle joint like a box corner, while a miter has reference to a joint such as is illustrated in Fig. 43, such as the corner of a picture frame.

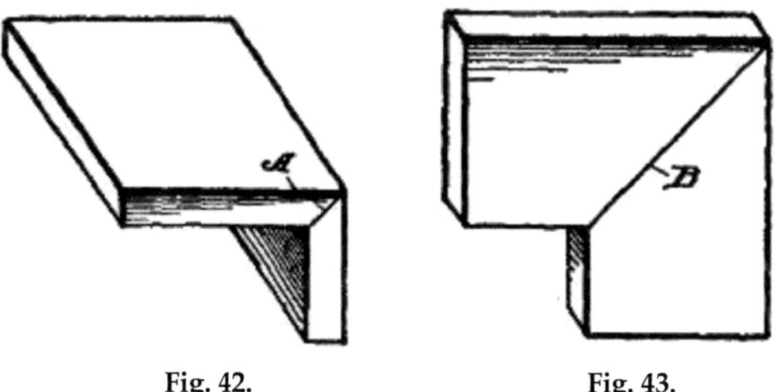

Fig. 42.　　　　　**Fig. 43.**

Proper Terms.—It is the application of the correct terms to things that lays the foundation for accurate thinking and proper expressions in describing work. A wise man once said that the basis of true science consists in correct definitions.

Picture Frames.—In picture frames the mitered corners may have a saw kerf (C) cut across the corners, as shown in Fig. 44, and a thin blade [Pg 54] of hard wood driven in, the whole being glued together.

Dovetail Joints.—It is in the laying out of the more complicated dovetail joints that the highest skill is required, because exactness is of more importance in this work than in any other article in joinery. In order to do this work accurately follow out the examples given, and you will soon be able to make a beautiful dovetail corner, and do it quickly.

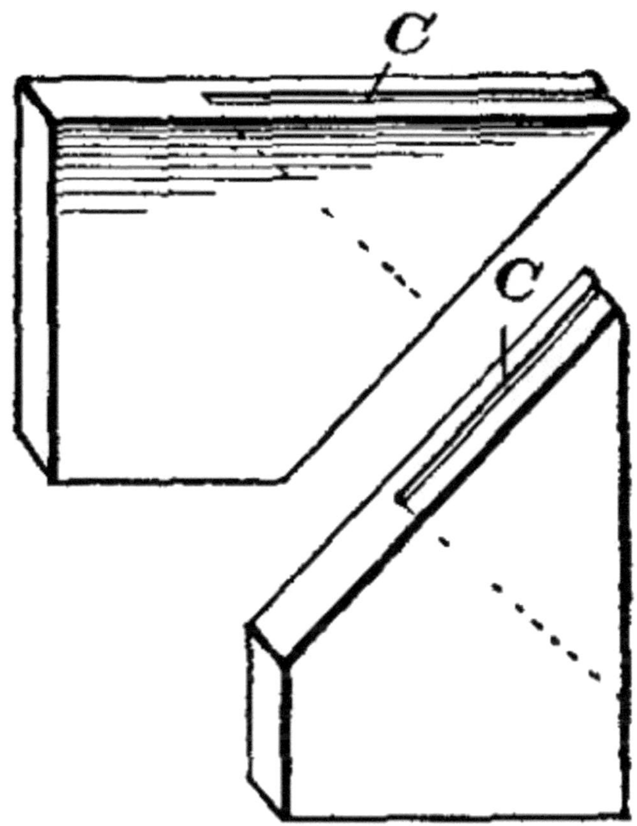

Fig. 44.

Preparing a Box Joint. — In order to match a box joint for the inner end of a table drawer, the first step is to select two work sides. One work side will be the edge of the board, and the other the side surface of the board, and on those surfaces we will put crosses, as heretofore suggested.

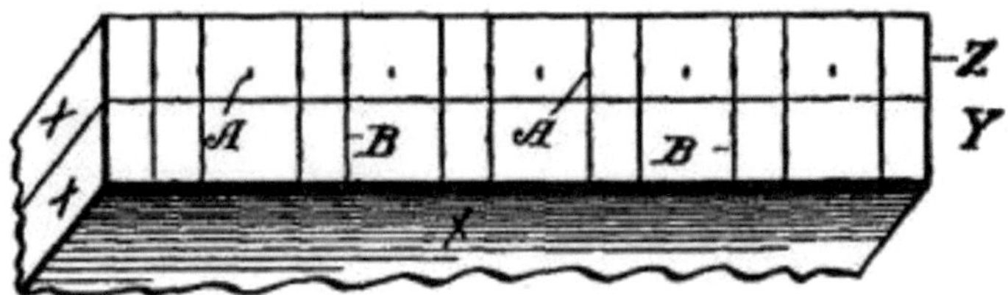

Fig. 45.

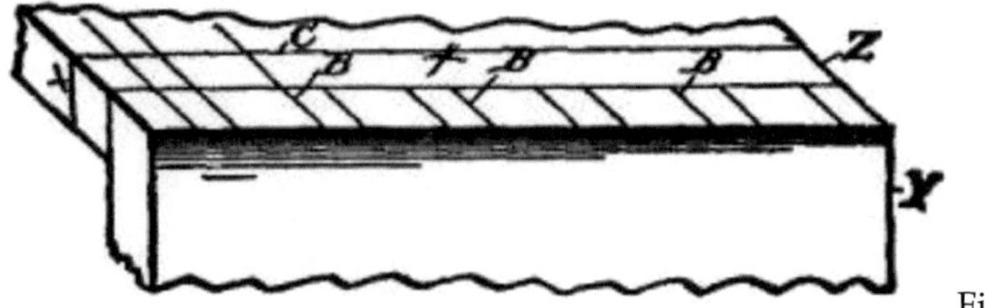

Fig. 46.

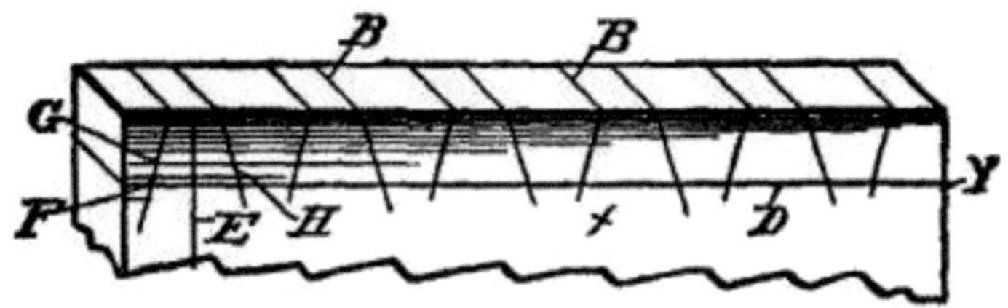

Fig. 47.

First Steps.—Now lap together the inner surfaces of these boards (Y, Z), so the ends are toward you, as shown in Fig. 45. Then, after measuring [Pg 55] the thickness of the boards to be joined (the thinnest, if they are of different thicknesses), set your compasses, or dividers, for ¼ inch, providing the boards are ½ inch thick, and, commencing at the work edge of the board, step off and point, as at A, the whole width of the board, and with a square make the two cross marks (B), using [Pg 56] the two first compass points (A), then skipping one, using the next two, and so on.

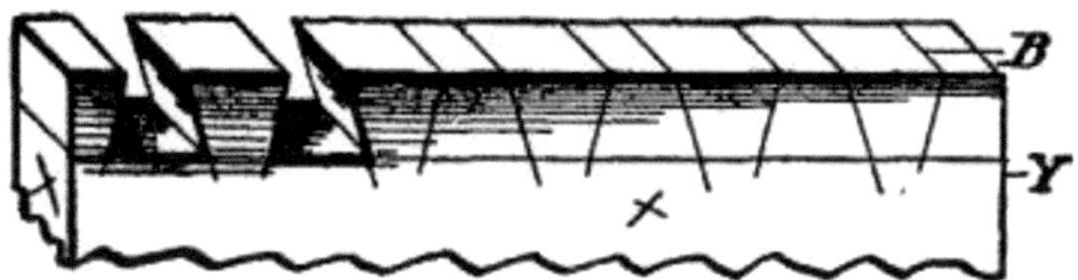

Fig. 48.

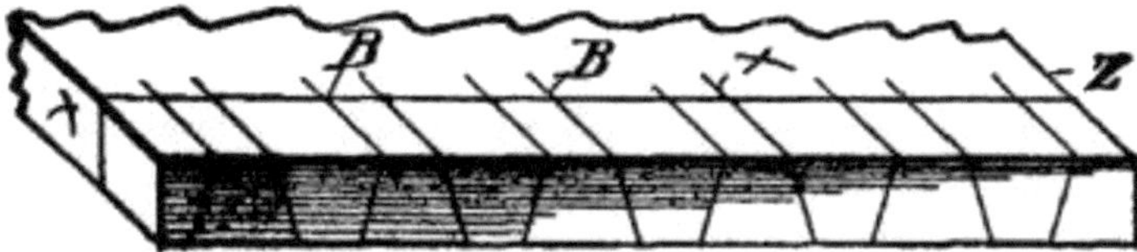

Fig. 49.

Fig. 50.

When this is done, turn up the board Z (Fig. 46), so that it is at right angles to the board Y, and so the outer surface of the board Z is flush with the end of the board X, and with a sharp knife point extend the lines B along with the grain of the wood on board Z,up to the cross mark C. This cross mark should have been previously made [Pg 57] and is located as far from the end of the board Z as the thickness of the board Y.

We now have the marks for the outer surface of the board Z, and the end marks of board Y. For the purpose of getting the angles of the end of the board Z and the outer side of board Y, a cross line (D, Fig. 47) is drawn across the board X near the end, this line being as far from the end as the thickness of the board Z, and a vertical line (E) is drawn midway between the two first cross marks (A).

Now, with your compass, which, in the meantime, has not been changed, make a mark (F), and draw down the line (G), which will give you the working angle at which you may set the bevel gage. Then draw down an angle from each alternate cross line (A), and turn the bevel and draw down the lines (H). These lines should all be produced on the opposite side of the board, so as to assure accuracy, and to this end the edges of the board also should be scribed.

Cutting Out the Spaces.—In cutting out the intervening spaces, which should be done with a sharp chisel, care should be observed not to cut over the shoulder lines. To prevent mistakes you should put some distinctive mark on each part to be cut away. In this instance E, H show the parts [Pg 58] to be removed, and in Fig. 48 two of the cutaway portions are indicated.

When the end of the board Z is turned up (Fig. 49), it has merely the longitudinal parallel lines B. The bevel square may now be used in the same manner as on the side of the board Y, and the fitting angles will then be accurately true.

This is shown in Fig. 50, in which, also, two of the cutaway parts are removed.

Tools Used in Laying Out Tenons and Mortises.—A sharp-pointed knife must always be used for making all marks. Never employ an awl for this work, as the fiber of the wood will be torn up by it. A small try square should always be used (not the large iron square), and this with a sharp-pointed compass and bevel square will enable you to turn out a satisfactory piece of work.

The foregoing examples, carefully studied, will enable you to gather the principles involved in laying off any work. If you can once make a presentable box joint, so that all the dovetails will accurately fit together, you will have accomplished one of the most difficult phases of the work, and it is an exercise which will amply repay you, because you will learn to appreciate what accuracy means.

[Pg 59]

CHAPTER VI

THE USES OF THE COMPASS AND THE SQUARE

The Square.—The square is, probably, the oldest of all tools, and that, together with the compass, or dividers, with which the square is always associated, has constituted the craftsman's emblem from the earliest historical times. So far as we now know, the plain flat form, which has at least one right angle and two or more straight edges, was the only form of square used by the workman. But modern uses, and the development of joinery and cabinet making, as well as the more advanced forms of machinery practice, necessitated new structural forms in the square, so that the bevel square, in which there is an adjustable blade set in a handle, was found necessary.

The Try Square.—In the use of the ordinary large metal square it is necessary to lay the short limb of the square on the face of the work, and the long limb must, therefore, rest against the work side or edge of the timber, so that the scribing edge of the short limb does not rest flat against the work. As such a tool is defective in work requiring accuracy, it brought into existence [Pg 60] what is called the try square, which has a rectangular handle, usually of wood, into which is fitted at one end a metal blade, which is at right angles to the edge of the handle. The handle, therefore, always serves as a guide for the blade in scribing work, because it lies flat down on the work.

The T-Square is another modification of the try square, its principal use being for draughting purposes.

The Compass.—The compass is one of the original carpenter's tools. The difference between *compass* and *dividers* is that compasses have adjustable pen or pencil points, whereas dividers are without adjustable points. Modern work has brought refinements in the character of the compass and dividers, so that we now have the bow-compass, which is, usually, a small tool, one leg of which carries a pen or pencil point, the two legs being secured together, usually, by a spring bow, or by a hinged joint with a spring attachment.

Proportional Dividers.—A useful tool is called the proportional dividers, the legs of which are hinged together intermediate the ends, so that the pivotal joint is adjustable. By means of this tool the scale of work may be changed, although its widest field of usefulness is work laid off on a [Pg 61] scale which you intend to reduce or enlarge proportionally.

Determining Angles.—Now, in order to lay out work the boy should know quickly and accurately how to determine various angles used or required in his work. The quickest way in which to learn this is to become familiar with the degree in its various relations.

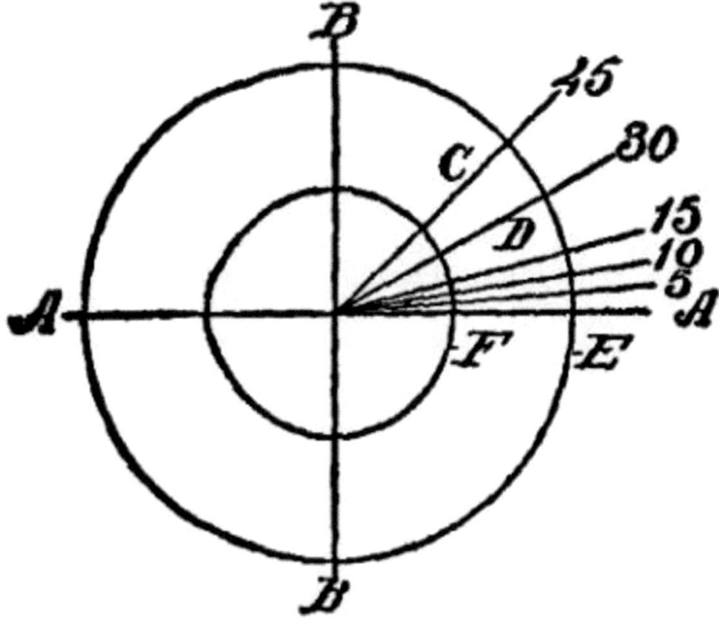

Fig. 51.

Definition of Degree.—A degree is not a measure, as we would designate a foot or a pound to determine distance or quantity. It is used to denote a division, space, interval or position. To illustrate, look at the circle, Fig. 51. The four cardinal points are formed by the cross lines (A, B), and in each one of the quadrants thus formed the circle is divided into 90 degrees. Look at the radial lines (C, D), and you will find that the distance between these lines is different along the [Pg 62] curved line (E) than along the curved line (F). The degree

68

is, therefore, to indicate only the space, division or interval in the circle.

The Most Important Angle.—Most important for one to know at a glance is that of 45 degrees, because the one can the more readily calculate the other degrees, approximately, by having 45 degrees once fixed in the mind, and impressed on the visual image. With a square and a compass it is a comparatively easy matter accurately to step off 45 degrees, as it is the line C, midway between A and B, and the other degrees may be calculated from the line C and the cardinal lines A or B.

Degrees Without a Compass.—But in the absence of a compass and when you do not wish to step off a circle, you will in such case lay down the square, and mark off at the outer margin of the limbs two equal dimensions. Suppose we take 2 inches on each limb of the square. The angle thus formed by the angle square blade is 45 degrees. To find 30 degrees allow the blade of the angle square to run from 2 inches on one limb to 3½ inches on the other limb, and it will be found that for 15 degrees the blade runs from 2 inches on one limb to 7½ inches on the other limb. It would be well to fix firmly these three points, at least, in your mind, as they will be of the utmost value to you. It is a comparatively easy matter now to [Pg 63] find 10 degrees or 25 degrees, or any intermediate line.

What Degrees Are Calculated From.—The question that now arises is what line one may use from which to calculate degrees, or at what point in the circle zero is placed. Degrees may be calculated either from the horizontal or from the vertical line. Examine Fig. 53. The working margin indicated by the cross mark is your base line, and in specifying an angle you calculate it from the work edge. Thus, the line A indicates an angle of 30 degrees. The dotted line is 45 degrees.

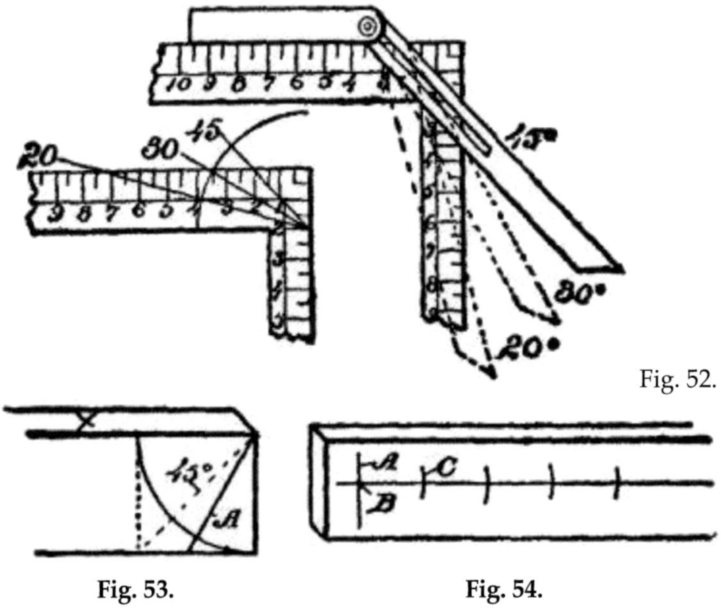

Fig. 52.

Fig. 53.   Fig. 54.

[Pg 64] The Dividers. — The dividers are used not only for scribing circles, but also for stepping and dividing spaces equally. There is a knack in the use of the dividers, where accuracy is wanted, and where the surface is of wood. Unless the utmost care is observed, the spaces will be unequal, for the reason that the point of the dividers will sink more deeply into the wood at some places than at others, due to the uneven texture of the wood grain. It will be better to make a line lengthwise, and a cross line (A) for starting (see Fig. 54). You may then insert one point of the dividers at the initial mark (B), and describe a small arc (C). Then move the dividers over to the intersection of the arc (C) on the line, and make the next mark, and so on.

Some useful hints along this same line will be found under the chapter on Drawing, which should be carefully studied.

[Pg 65]

CHAPTER VII

HOW THE DIFFERENT STRUCTURAL PARTS ARE DESIGNATED

The Right Name for Everything.—Always make it a point to apply the right term to each article or portion of a structure. Your explanation, to those who do know the proper technical terms, will render much easier a thorough understanding; and to those who do not know, your language will be in the nature of an education.

Proper Designations.—Every part in mechanism, every point, curve and angle has its peculiar designation. A knowledge of terms is an indication of thoroughness in education, and, as heretofore stated, becomes really the basis of art, as well as of the sciences. When you wish to impart information to another you must do it in terms understood by both.

Furthermore, and for this very reason, you should study to find out how to explain or to define the terms. You may have a mental picture of the structure in your mind, but when asked to explain it you are lost.

Learning Mechanical Forms.—Suppose, for example, we take the words *segment* and *sector*. [Pg 66] Without a thorough understanding in your own mind you are likely to confuse these terms by taking one for the other. But let us assume you are to be called upon to explain a sector to some one who has no idea of terms and their definitions. How would you describe it? While it is true it is wedge-shaped, you will see by examining the drawing that it is not like a wedge. The sector has two sides running from a point like a wedge, but the large end of the sector is curved.

If you were called upon to define a segment you might say it had one straight line and one curve, but this would not define it very lucidly. Therefore, in going over the designations given, not only fix in your mind the particular form, but try to remember some particular manner in which you can clearly express the form, the shape or the relation of the parts.

For your guidance, therefore, I have given, as far as possible, simple figures to aid you in becoming acquainted with structures and their designations, without repeating the more simple forms which I have used in the preceding chapters.

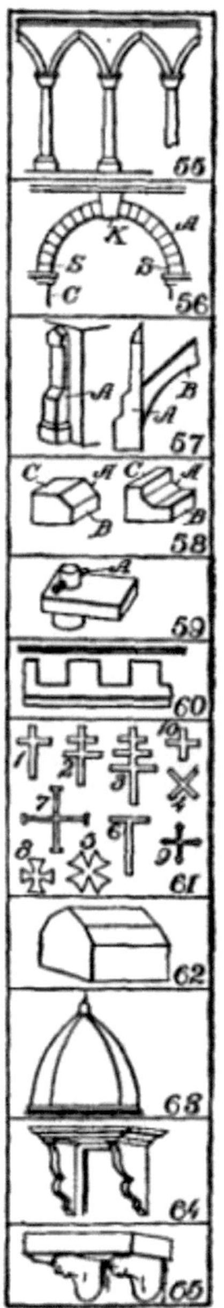

55. *Arcade.*—A series of arches with the columns or piers which support them, the spandrels above, and other parts.

[Pg 67] 56. *Arch.*—A curved member made up, usually, of separate wedge-shaped solids, A. K, Keystone; S, Springers; C, Chord, or span.

57. *Buttress.*—A projecting mass of masonry. A, used for resisting the thrust of an arch, or for ornamentation; B, a flying buttress.

58. *Chamfer.*—The surface A formed by cutting away the arris or angle formed by two faces, B, C, of material.

59. *Cotter or Cotter Pin.*—A pin, A, either flat, square or round, driven through a projecting tongue to hold it in position.

60. *Crenelated.*—A form of molding indented or notched, either regularly or irregularly.

61. *Crosses.*—1. Latin cross, in the Church of Rome carried before Bishops. 2. Double cross, carried before Cardinals and Bishops. 3. Triple or Papal cross. 4. St. Andrew's and St. Peter's cross. 5. Maltese cross. 6. St. Anthony or Egyptian cross. 7. Cross of Jerusalem. 8. A cross patté or fermé (head or first). 9. A cross patonce (that is, growing larger at the ends). 10. Greek cross.

62. *Curb Roof.*—A roof having a double slope, or composed on each side of two parts which have unequal inclinations; a gambrel roof.

63. *Cupola.*—So called on account of its resemblance to a cup. A roof having a rounded form. When on a large scale it is called a dome.

Crown Post.—See *King Post.*

64. *Console.*—A bracket with a projection not more than half its height.

65. *Corbels.*—A mass of brackets to support a shelf or structure. Largely employed in Gothic architecture.

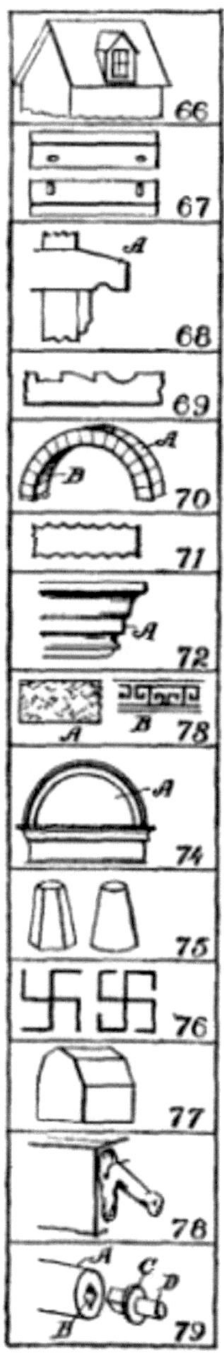

66. Dormer.—A window pierced in a roof and so set as to be vertical, while the roof slopes away from it. Also called a Gablet.

67. Dowel.—A pin or stud in one block, or body, designed to engage with holes in another body to hold them together in alignment.

68. Drip.—That part of a cornice or sill course A, or other horizontal member which projects beyond the rest, so as to divert water.

[Pg 68] 69. Detents.—Recesses to lock or to serve as a stop or holding place.

70. Extrados.—The exterior curve of an arch, especially the upper curved face A. B is the Intrados or Soffit.

71. Engrailed.—Indented with small concave curves, as the edge of a bordure, bend, or the like.

72. Facet.—The narrow plain surface, as A, between the fluting of a column.

73. Fret, Fretwork.—Ornamental work consisting of small fillets, or slats, intersecting each other or bent at right angles. Openwork in relief, when elaborated and minute in all its parts. Hence any minute play of light and shade. A, Japanese fretwork. B, Green fret.

74. Frontal, also called Pediment.—The triangular space, A, above a door or window.

75. Frustums.—That part of a solid next the base, formed by cutting off the top; or the part of any solid, as of a cone, pyramid, etc., between two planes, which may either be parallel or inclined to each other.

76. Fylfat.—A rebated cross used as a secret emblem and worn as an ornament. It is also called Gammadium, and more commonly known as Swastika.

77. Gambrel Roof.—A curb roof having the same section in all its parts, with a lower, steeper and longer part. See Curb Roof and distinguish difference.

78. Gargoyle.—A spout projecting from the roof gutter of a building, often carved grotesquely.

79. Gudgeon. — A wooden shaft, A, with a socket, B, into which is fitted a casting, C. The casting has a gudgeon, D.

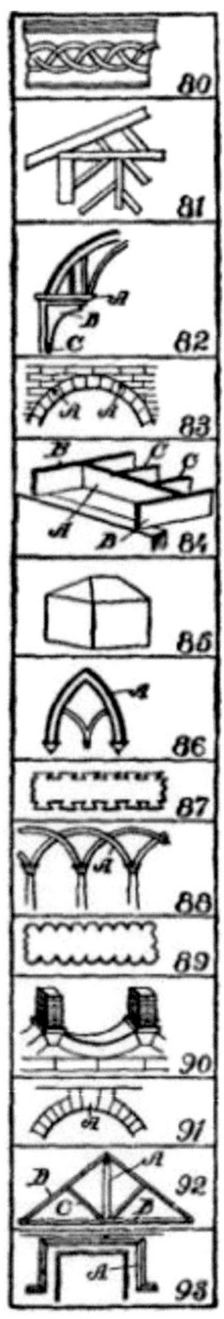

80
81
82
A
Д
С
83
Л Л
84
В С С
А D
85
86
A
87
88
A
89
90
91
Л
92
А
В
С D
93
А

80. Guilloche.—An ornament in the form of two or more bands or strings twisted together or over or through each other.

81. Half Timbered.—Constructed of a timber frame, having the spaces filled in with masonry.

82. Hammer Beam.—A member of one description of roof truss, called hammer-beam truss, which is so framed as not to have a tie beam [Pg 69] at the top of the wall. A is the hammer beam, and C the pendant post.

83. Haunches.—The parts A, A, on each side of the crown of an arch. Each haunch is from one-half to two-thirds of the half arch.

84. Header.—A piece of timber, A, fitted between two trimmers, B, B, to hold the ends of the tail beams, C, C.

85. Hip Roof.—The external angle formed by the meeting of two sloping sides or skirts of a roof which have their wall plates running in different directions.

86. Hood Molding.—A projecting molding over the head of an arch, as at A, forming the outer-most member of the archivolt.

87. Inclave.—The border, or borders, having a series of dovetails. One variation of molding or ornamentation.

88. Interlacing Arch.—Arches, usually circular, so constructed that their archivolts, A, intersect and seem to be interlaced.

89. Invected.—Having a border or outline composed of semicircles or arches, with the convexity outward. The opposite of engrailed.

90. Inverted Arch.—An arch placed with the crown downward; used in foundation work.

91. Keystone.—The central or topmost stone, A, of an arch, sometimes decorated with a carving.

92. King Post.—A member, A, of a common form of truss for roofs. It is strictly a tie intended to prevent the sagging of the tie beam, B, in the middle. If there are struts, C, supporting the rafters, D, they extend down to the foot of the King Post.

93. Label. — The name given to the projecting molding, A, around the top of the door opening. A form of mediæval architecture.

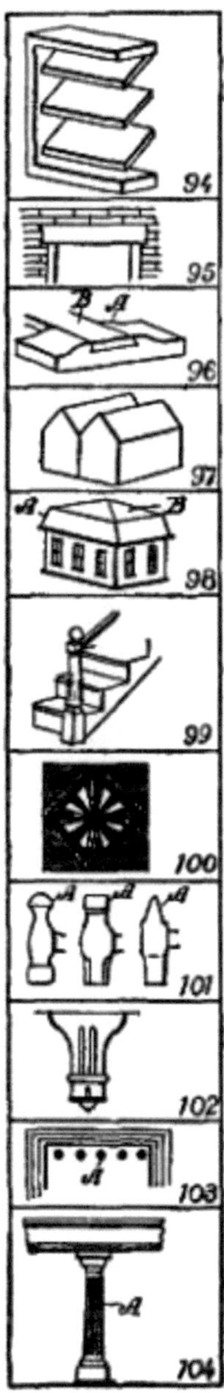

94
95
B A
96
97
A B
98
99
100
A A A
101
102
A
103
A
104

94. Louver.—The sloping boards, A, set to shed rain water outward in an opening of a frame, as in belfry windows.

[Pg 70] 95. Lintel.—A horizontal member. A spanning or opening of a frame, and designed to carry the wall above it.

96. Lug.—A. projecting piece, as A, to which anything is attached, or against which another part, like B, is held.

97. M-Roof.—A kind of roof formed by the junction of two common roofs with a valley between them, so the section resembles the letter M.

98. Mansard Roof.—A hipped curb roof, that is, a roof having on all sides two slopes, the lower one, A, being steeper than the upper portion or deck.

99. Newel Post.—The upright post at the foot of a stairway, to which the railing is attached.

100. Parquetry.—A species of joinery or cabinet work, consisting of an inlay of geometric or other patterns, generally of different colored woods, used particularly for floors.

101. Peen. also Pein.—The round, round-edged or hemispherical end, as at A, of a hammer.

102. Pendant.—A hanging ornament on roofs, ceilings, etc., and much used in the later styles of Gothic architecture where it is of stone. Imitated largely in wood and plaster work.

103. Pentastyle.—A pillar. A portico having five pillars, A, is called the Pentastyle in temples of classical construction.

104. Pedestal.—An upright architectural member, A, right-angled in plan, constructionally a pier, but resembling a column, having a capital, shaft and base to agree with the columns in the structure.

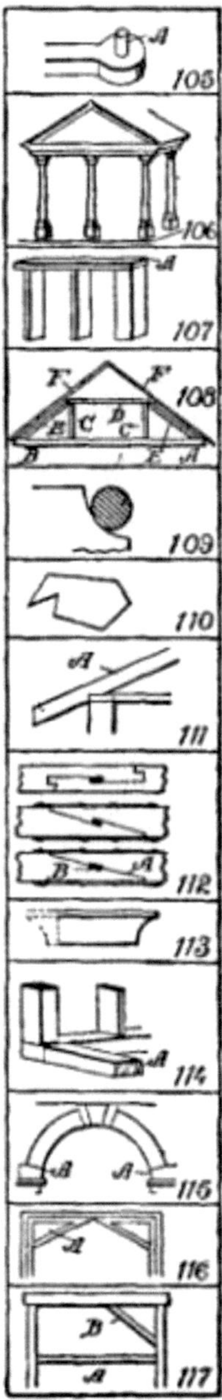

105
106
107
108
109
110
111
112
113
114
115
116
117

105. Pintle.—An upright pivot pin, or the pin of a hinge; A represents the pintle of a rudder.

106. Portico.—A colonnade or covered structure, especially in classical style, of architecture, and usually at the entrance of a building.

107. Plate.—A horizontal timber, A, used as a top or header for supporting timbers, roofs and the like.

[Pg 71] 108. Queen Post.—One of two suspending posts in a roof truss, or other framed truss of simple form. Compare with King Post. A, B, tie beam; C, C, queen posts; D, straining piece; E, principal rafter; F, rafter.

109. Quirk Molding.—A small channel, deeply recessed, in proportion to its width, used to insulate and give relief to a convex rounded molding. An excellent corner post for furniture.

110. Re-entering.—The figure shows an irregular polygon (that is, many-sided figure) and is a re-entering polygon. The recess A is a re-entering angle.

111. Rafter.—Originally any rough and heavy piece of timber, but in modern carpentry used to designate the main roof support, as at A. See Queen Post.

112. Scarfing.—Cutting timber at an angle along its length, as the line A. Scarfing joints are variously made. The overlapping joints may be straight or recessed and provided with a key block B. When fitted together they are securely held by plates and bolts.

113. Scotia Molding.—A sunken molding in the base of a pillar, so called from the dark shadow which it casts.

114. Sill.—In carpentry the base piece, or pieces, A, on which the posts of a structure are set.

115. Skew-Back.—The course of masonry, such as a stone, A, with an inclined face, which forms the abutment for the voussoirs, B, or wedge-shaped stones comprising the arch.

116. Spandrel.—The irregular, triangular space, A, between the curve of an arch and the enclosing right angle.

117. Strut. — In general, any piece of a frame, such as a timber A, or a brace B, which resists pressure or thrust in the direction of its length.

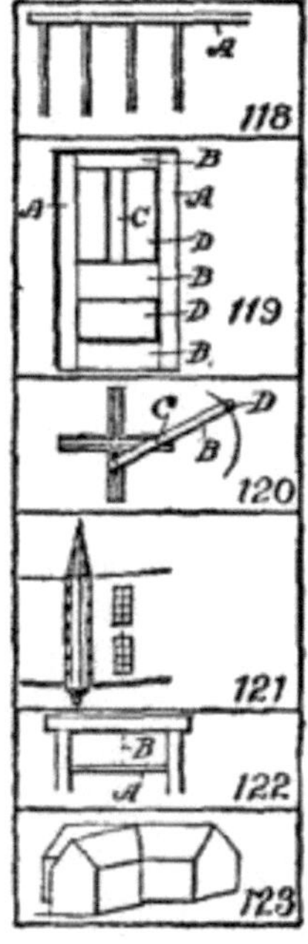

118. Stud, Studding. — The vertical timber or scantling, A, which is one of the small uprights of a building to which the boarding or plastering lath are nailed.

[Pg 72] 119. Stile. — The main uprights of a door, as A, A; B, B, B, rails; C, C, mullions; D, D, panels.

Tie Beam. — See Queen Post.

120. Trammel. — A very useful tool for drawing ellipses. It comprises a cross, A, with grooves and a bar, B, with pins, C, attached to sliding blocks in the grooves, and a pen or stylus, D, at the projecting end of the bar to scribe the ellipse.

121. Turret. — A little tower, frequently only an ornamental structure at one of the angles of a larger structure.

122. Transom.—A horizontal cross-bar, A, above a door or window or between a door and a window above it. Transom is the horizontal member, and if there is a vertical, like the dotted line B, it is called a Mullion. See Stile.

123. Valley Roof.—A place of meeting of two slopes of a roof which have their sides running in different directions and formed on the plan of a re-entrant angle.

[Pg 73]

CHAPTER VIII

DRAWING AND ITS UTILITY

A knowledge of drawing, at least so far as the fundamentals are concerned, is of great service to the beginner. All work, after being conceived in the brain, should be transferred to paper. A habit of this kind becomes a pleasure, and, if carried out persistently, will prove a source of profit. The boy with a bow pen can easily draw circles, and with a drawing or ruling pen he can make straight lines.

Representing Objects.—But let him try to represent some object, and the pens become useless. There is a vast difference in the use of drawing tools and free-hand drawing. While the boy who is able to execute free-hand sketches may become the better artist, still that art would not be of much service to him as a carpenter. First, because the use of tools gives precision, and this is necessary to the builder; and, second, because the artist deals wholly with perspectives, whereas the builder must execute from plane surfaces or elevations.

Forming Lines and Shadows.—It is not my intention to furnish a complete treatise on this subject, [Pg 74] but to do two things, one of which will be to show, among other features, how simple lines form objects; how shading becomes an effective aid; how proportions are formed; and, second, how to make irregular forms, and how they may readily be executed so that the boy may be able to grasp the ideas for all shapes and structural devices.

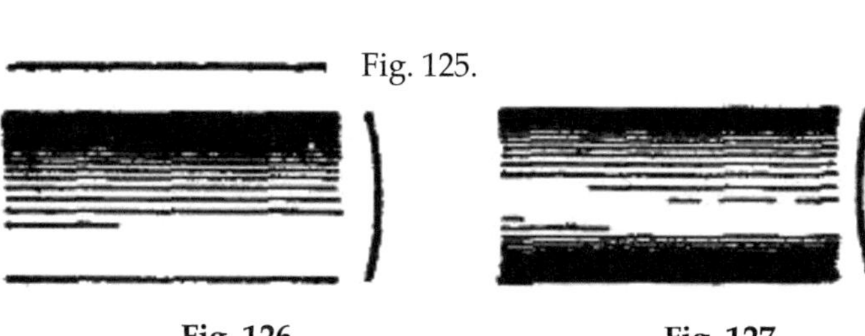

Fig. 125.

Fig. 126.   **Fig. 127.**

Analysis of Line Shading.—In the demonstration of this work I shall give an analysis of the simple lines formed, showing the terms used to designate the lines, curves, and formations, so that when any work is laid out the beginner will be able, with this glossary before him, to describe architecturally, as well as mathematically, the angles and curves with which he is working.

How to Characterize Surface.—Suppose we commence simply with straight lines. How shall [Pg 75] we determine the character of the surface of the material between the two straight lines shown in Fig. 125? Is it flat, rounded, or concaved? Let us see how we may treat the surface by simple lines so as to indicate the configuration.

Fig. 128.   Fig. 129.

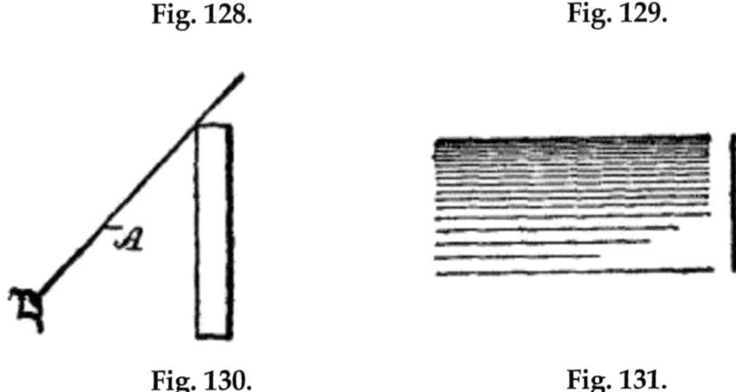

Fig. 130.   Fig. 131.

Concave Surfaces. — In Fig. 126 the shading lines commence at the upper margin, and are heaviest there, the lines gradually growing thinner and farther apart.

Convex Surfaces. — In Fig. 127 the shading is very light along the upper margin, and heavy at the lower margin. The first shaded figure, therefore, represents a concaved surface, and the second [Pg 76] figure a convex surface. But why? Simply for the reason that in drawings, as well as in nature, light is projected downwardly, hence when a beam of light moves past the margin of an object, the contrast at the upper part, where the light is most intense, is strongest.

The shading of the S-shaped surface (Fig. 128) is a compound of Figs. 126 and 127.

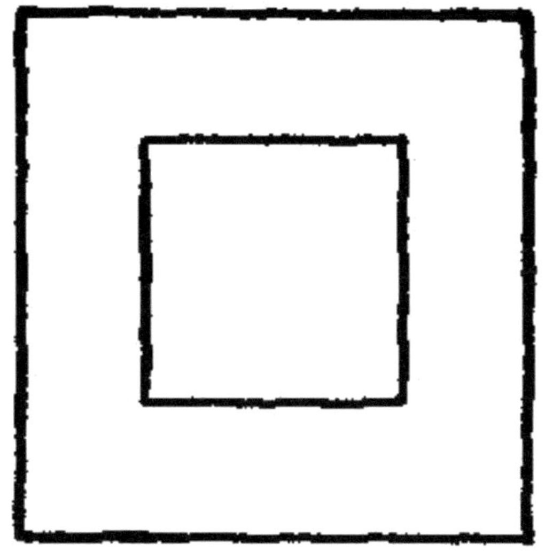

Fig. 132.

Shadows From a Solid Body.—We can understand this better by examining Fig. 129, which shows a vertical board, and a beam of light (A) passing downwardly beyond the upper margin of the board. Under these conditions the upper margin of the board appears darker to the vision, by contrast, than the lower part. It should also be understood that, in general, the nearer the object the lighter it is, so that as the upper edge of the board is farthest from the eye the heavy shading there will at least give the appearance of distance to that edge.

[Pg 77] But suppose that instead of having the surface of the board flat, it should be concaved, as in Fig. 130, it is obvious that the hollow, or the concaved, portion of the board must intensify the shadows or the darkness at the upper edge. This explains why the heavy shading in Fig. 126 is at that upper margin.

Flat Effects.—If the board is flat it may be shaded, as shown in Fig. 131, in which the lines are all of the same thickness, and are spaced farther and farther apart at regularly increasing intervals.

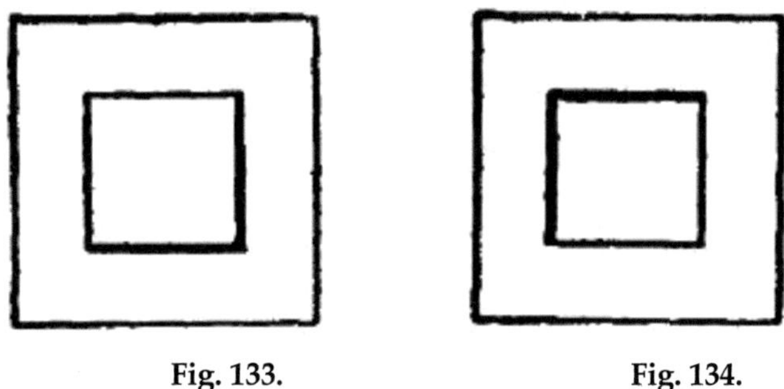

Fig. 133. Fig. 134.

The Direction of Light.—Now, in drawing, we must observe another thing. Not only does the light always come from above, but it comes also from the left side. I show in Fig. 132 two squares, one within the other. All the lines are of the same thickness. Can you determine by means of such a drawing what the inner square represents? Is it a block, or raised surface, or is it a depression?

[Pg 78] Raised Surfaces.—Fig. 133 shows it in the form of a block, simply by thickening the lower and the right-hand lines.

Depressed Surfaces.—If, by chance, you should make the upper and the left-hand lines heavy, as in Fig. 134, it would, undoubtedly, appear depressed, and would need no further explanation.

Full Shading,—But, in order to furnish an additional example of the effect of shading, suppose we shade the surface of the large square, as shown in Fig. 135, and you will at once see that not only is the effect emphasized, but it all the more clearly expresses what you want to show. In like manner, in Fig. 136, we shade only the space within the inner square, and it is only too obvious how shadows give us surface conformation.

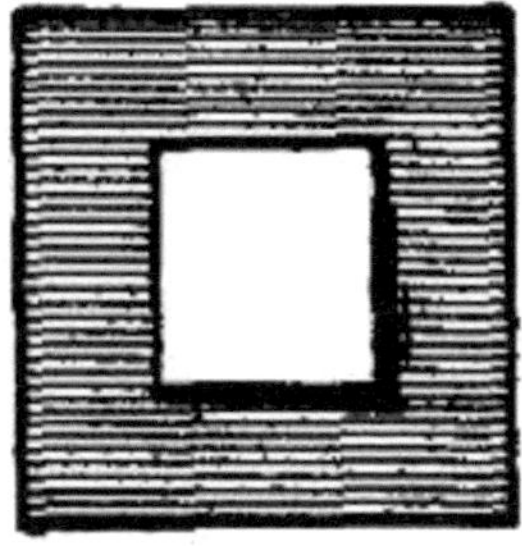

Fig. 135.

Fig. 136.

Illustrating Cube Shading.—In Fig. 137 I show merely nine lines joined together, all lines being of equal thickness.

As thus drawn it may represent, for instance, [Pg 79] a cube, or it may show simply a square base (A) with two sides (B, B) of equal dimensions.

Shading Effects.—Now, to examine it properly so as to observe what the draughtsman wishes to express, look at Fig. 138, in which the three diverging lines (A, B, C) are increased in thickness, and the cube appears plainly. On the other hand, in Fig. 139, the thickening of the lines (D, E, F) shows an entirely different structure.

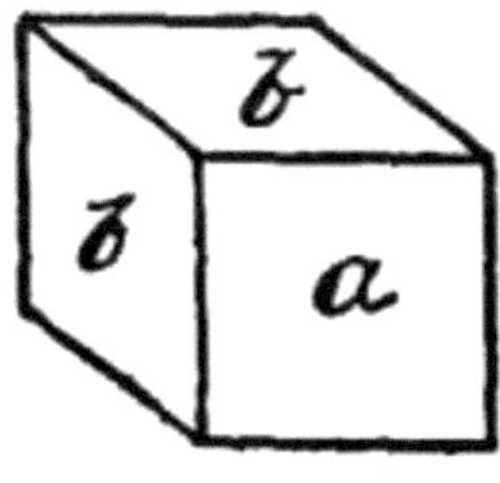

Fig. 137.

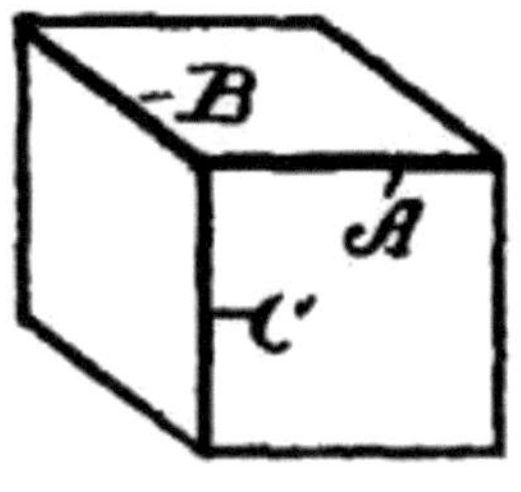
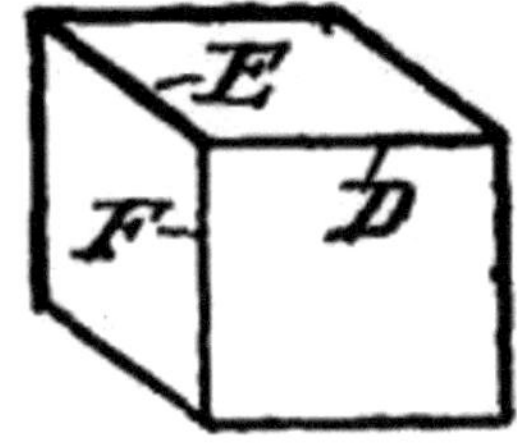

Fig. 138.   **Fig. 139.**

It must be remembered, therefore, that to show raised surfaces the general direction is to shade heavily the lower horizontal and the right vertical lines. (See Fig. 133.)

Heavy Lines.—But there is an exception to this rule. See two examples (Fig. 140). Here two parallel [Pg 80] lines appear close together to form the edge nearest the eye. In such cases the second, or upper, line is heaviest. On vertical lines, as in Fig. 141, the second line from the right is heaviest. These examples show plain geometrical lines, and those from Figs. 138 to 141, inclusive, are in perspective.

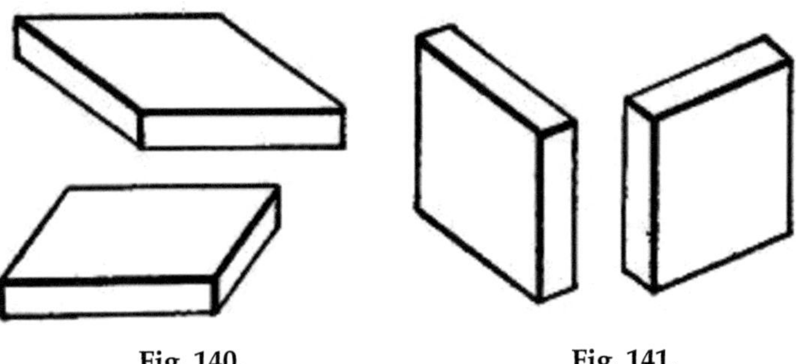

Fig. 140.   **Fig. 141.**

Perspective.—A perspective is a most deceptive figure, and a cube, for instance, may be drawn so that the various lines will differ in length, and also be equidistant from each other. Or all the lines may be of the same length and have the distances between them vary. Supposing we have two cubes, one located above the other, separated, say, two feet or more from each other. It is obvious that the lines of the two cubes will not be the same to a camera, because, if they were photographed, they would appear exactly as they are, so far as their positions are concerned, and not as they appear. But the cubes do appear to the eye [Pg 81] as having six equal sides. The camera shows that they do not have six equal sides so far as measurement is concerned. You will see, therefore, that the position of the eye, relative to the cube, is what determines the angle, or **the relative** angles of all the lines.

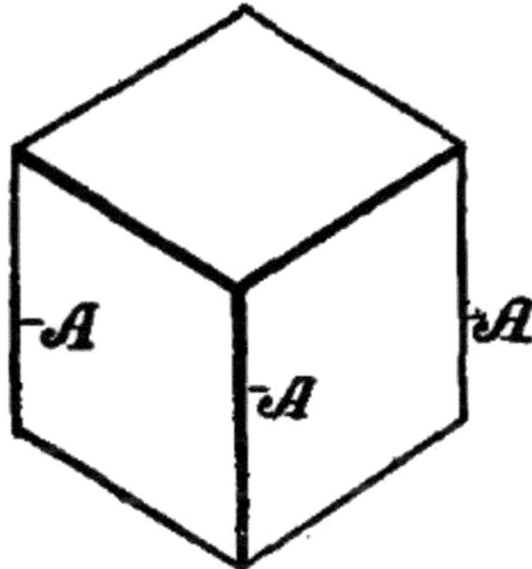

Fig. 142.

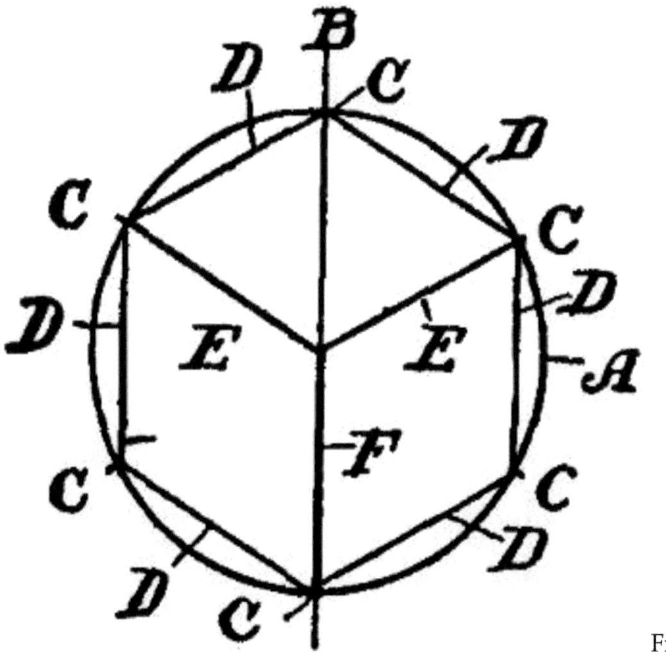

Fig. 143.

A True Perspective of a Cube.—Fig. 142 shows a true perspective—that is, it is true from the measurement standpoint. It is what is called an *isometrical* view, or a figure in which all the lines not only are of equal length, but the parallel lines are [Pg 82] all spaced apart the same distances from each other.

Isometric Cube.—I enclose this cube within a circle, as in Fig. 143. To form this cube the circle (A) is drawn and bisected with a vertical line (B). This forms the starting point for stepping off the six points (C) in the circle, using the dividers without resetting, after you have made the circle. Then connect each of the points (C) by straight lines (D). These lines are called chords. From the center draw two lines (E) at an angle and one line (F) vertically. These are the radial lines. You will see from the foregoing that the chords (D) form the outline of the cube—or the lines farthest from the eye, and the radial lines (E, F) are the nearest to the eye. In this position we are looking at the block at a true diagonal—that is, from a corner at one side to the extreme corner on the opposite side.

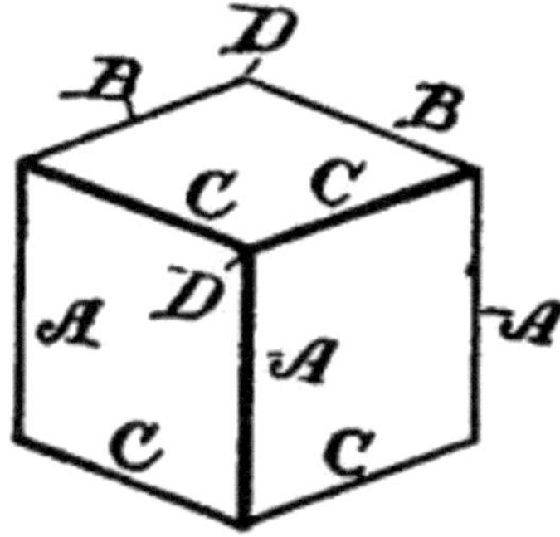

Fig. 144.

Let us contrast this, and particularly Fig. 142, [Pg 83] with the cube which is placed higher up, viewed from the same standpoint.

Flattened Perspective.—Fig. 144 shows the new perspective, in which the three vertical lines (A, A, A) are of equal length, and the six angularly disposed lines (B, C) are of equal length, but shorter than the lines A. The only change which has been made is to shorten the distance across the corner from D to D, but the vertical lines (A) are the same in length as the corresponding lines in Fig. 143. Notwithstanding this change the cubes in both figures appear to be of the same size, as, in fact, they really are.

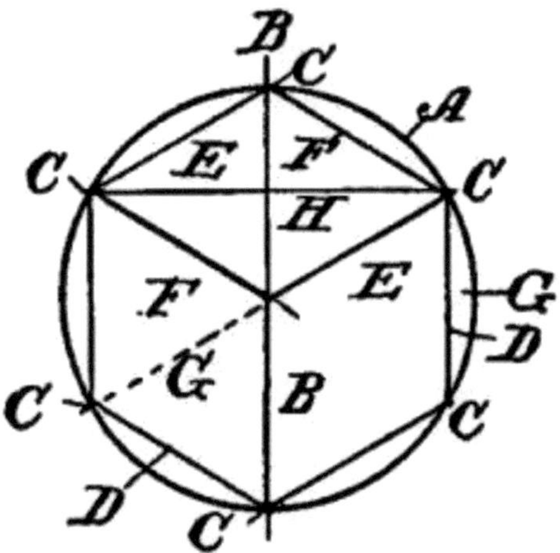

Fig. 145.

In forming a perspective, therefore, it would be a good idea for the boy to have a cube of wood always at hand, which, if laid down on a horizontal support, alongside, or within range of the object to [Pg 84] be drawn, will serve as a guide to the perspective.

Technical Designations.—As all geometrical lines have designations, I have incorporated such figures as will be most serviceable to the boy, each figure being accompanied by its proper definition.

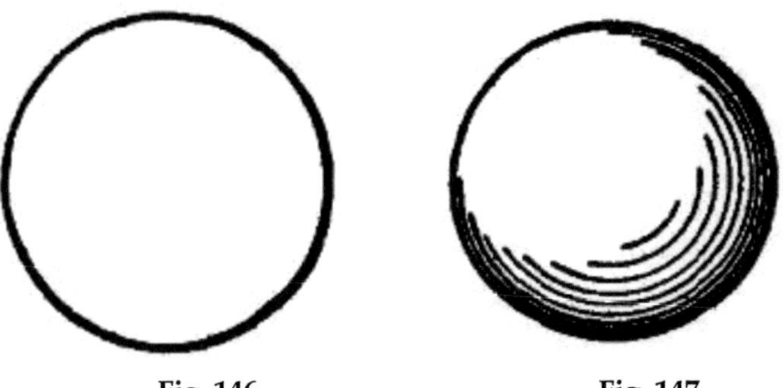

Fig. 146.   Fig. 147.

Before passing to that subject I can better show some of the simple forms by means of suitable diagrams.

Referring to Fig. 145, let us direct our attention to the body (G), formed by the line (D) across the circle. This body is called a segment. A chord (D) and a curve comprise a segment.

Sector and Segment.—Now examine the shape of the body formed by two of the radial lines (E, E) and that part of the circle which extends from one radial line to the other. The body thus formed is a sector, and it is made by two radiating lines and a curved line. Learn to distinguish readily, in your mind, the difference between the two figures.

[Pg 85] Terms of Angles.—The relation of the lines to each other, the manner in which they are joined together, and their comparative angles, all have special terms and meanings. Thus, referring to the isometric cube, in Fig. 145, the angle formed at the center by the lines (B, E) is different from the angle formed at the margin by the lines (E, F). The angle formed by B, E is called an exterior angle; and that formed by E, F is an interior angle. If you will draw a line (G) from the center to the circle line, so it intersects it at C, the lines B, D, G form an equilateral or isosceles triangle; if you draw a chord (A) from C to C, the lines H, E, F will form an obtuse triangle, and B, F, H a right-angled triangle.

Circles and Curves.—Circles, and, in fact, all forms of curved work, are the most difficult for beginners. The simplest figure is the circle, which, if it represents a raised surface, is provided with a heavy line on the lower right-hand side, as in Fig. 146; but the proper artistic expression is shown in Fig. 147, in which the lower right-hand side is shaded in rings running only a part of the way around, gradually diminishing in length, and spaced farther and farther apart as you approach the center, thus giving the appearance of a sphere.

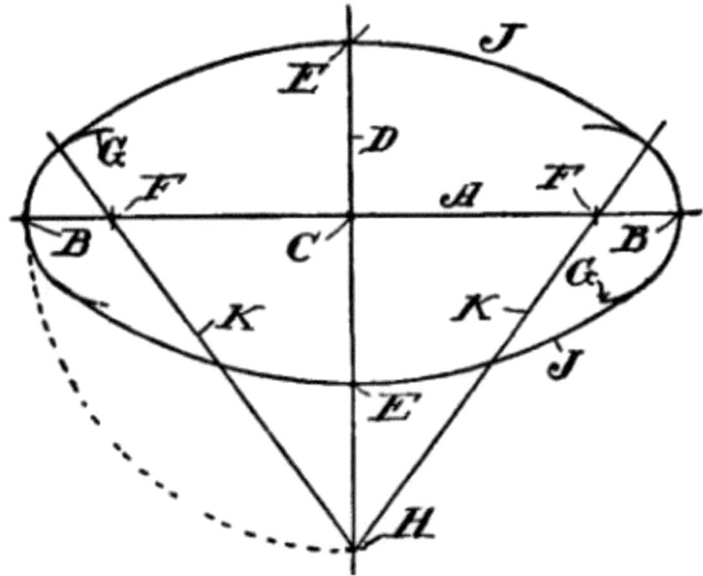

Fig. 148.

Irregular Curves. — But the irregular curves require the most care to form properly. Let us try [Pg 86] first the elliptical curve (Fig. 148). The proper thing is, first, to draw a line (A), which is called the "major axis." On this axis we mark for our guidance two points (B, B). With the dividers find a point (C) exactly midway, and draw a cross line (D). This is called the "minor axis." If we choose to do so we may indicate two points (E, E) on the minor axis, which, in this case, for convenience, are so spaced that the distance along the major axis, between B, B, is twice the length across the minor axis (D), along E, E. Now find one-quarter of the distance from B to C, as at F, and with a compass pencil make a half circle (G). If, now, you will set the compass point on the center mark (C), and the pencil point of the compass on B, and measure along the minor axis (D) on both [Pg 87] sides of the major axis, you will make two points, as at H. These points are your centers for scribing the long sides of the el-lipse. Before proceeding to strike the curved lines (J), draw a diago-nal line (K) from H to each marking point (F). Do this on both sides of the major axis, and produce these lines so they cross the curved

98

lines (G). When you ink in your ellipse do not allow the circle pen to cross the lines (K), and you will have a mechanical ellipse.

Ellipses and Ovals.—It is not necessary to measure the centering points (F) at certain specified distances from the intersection of the horizontal and vertical lines. We may take any point along the major axis, as shown, for instance, in Fig. 149. Let B be this point, taken at random. Then describe the half circle (C). We may, also, arbitrarily, take any point, as, for instance, D on the minor axis E, and by drawing the diagonal lines (F) we find marks on the circle (C), which are the meeting lines for the large curve (H), with the small curve (C). In this case we have formed an ovate or an oval form. Experience will soon make perfect in following out these directions.

Focal Points.—The focal point of a circle is its center, and is called the *focus*. But an ellipse has two focal points, called *foci*, represented by F, F in Fig. 148, and by B, B in Fig. 149.

[Pg 88] A *produced line* is one which extends out beyond the marking point. Thus in Fig. 148 that part of the line K between F and G represents the produced portion of line K.

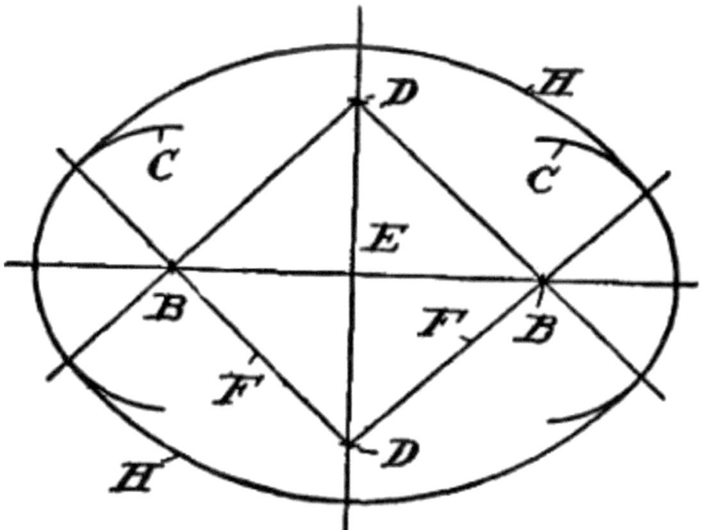

Fig. 149.

Spirals.—There is no more difficult figure to make with a bow or a circle pen than a spiral. In Fig. 150 a horizontal and a vertical line (A, B), respectively, are drawn, and at their intersection a small circle (C) is formed. This now provides for four centering points for the circle pen, on the two lines (A, B). Intermediate these points indicate a second set of marks halfway between the marks on the lines. If you will now set the point of the compass at, say, the mark 3, and the pencil point of the compass at D, and make a curved mark one-eighth of the way around, say, to the radial line (E), then put the point of the [Pg 89] compass to 4, and extend the pencil point of the compass so it coincides with the curved line just drawn, and then again make another curve, one-eighth of a complete circle, and so on around the entire circle of marking points, successively, you will produce a spiral, which, although not absolutely accurate, is the nearest approach with a circle pen. To make this neatly re-quires care and patience.

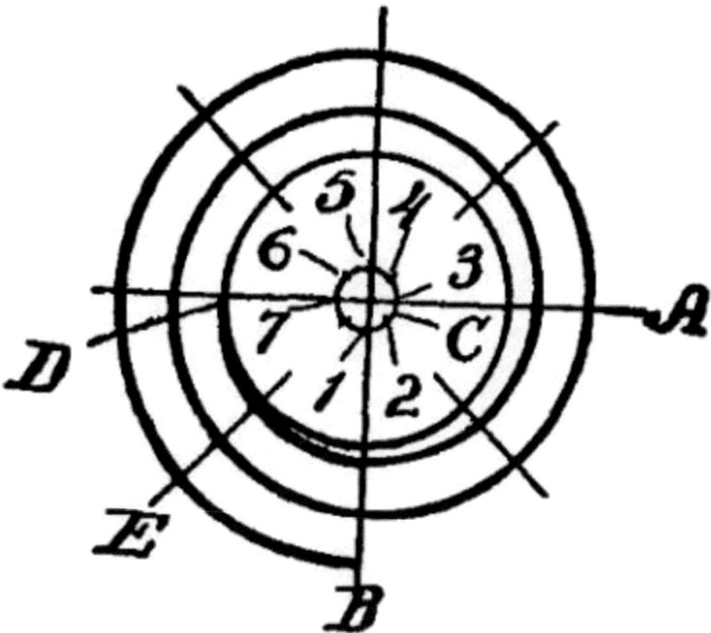

Fig. 150.

100

Perpendicular and Vertical. — A few words now as to terms. The boy is often confused in determining the difference between *perpendicular* and *vertical.* There is a pronounced difference. Vertical means up and down. It is on a line in the direction a ball takes when it falls straight toward the center of the earth. The word *perpendicular*, as usually employed in astronomy, means the same thing, but in geometry, or in drafting, or in its use in the arts it means that a perpendicular [Pg 90] line is at right angles to some other line. Suppose you put a square upon a roof so that one leg of the square extends up and down on the roof, and the other leg projects outwardly from the roof. In this case the projecting leg is *perpendicular* to the roof. Never use the word *vertical* in this connection.

Signs to Indicate Measurements. — The small circle (°) is always used to designate *degree.* Thus 10° means ten degrees.

Feet are indicated by the single mark '; and two closely allied marks " are for inches. Thus five feet ten inches should be written 5' 10". A large cross (×) indicates the word "by," and in expressing the term six feet by three feet two inches, it should be written 6' × 3'2".

The foregoing figures give some of the fundamentals necessary to be acquired, and it may be said that if the boy will learn the principles involved in the drawings he will have no difficulty in producing intelligible work; but as this is not a treatise on drawing we cannot go into the more refined phases of the subject.

Definitions. — The following figures show the various geometrical forms and their definitions:

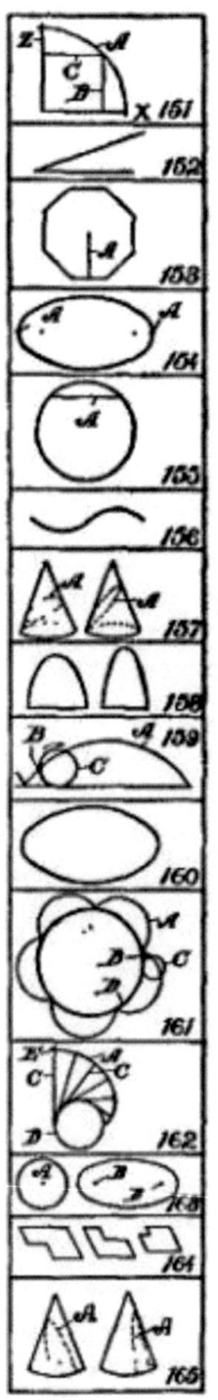

151. *Abscissa.*—The point in a curve, A, which is referred to by certain lines, such as B, which extend out from an axis, X, or the ordinate line Z.

[Pg 91] 152. *Angle.*—The inclosed space near the point where two lines meet.

153. *Apothegm.*—The perpendicular line A from the center to one side of a regular polygon. It represents the radial line of a polygon the same as the radius represents half the diameter of a circle.

154. *Apsides* or *Apsis.*—One of two points, A, A, of an orbit, oval or ellipse farthest from the axis, or the two small dots.

155. *Chord.*—A right line, as A, uniting the extremities of the arc of a circle or a curve.

156. *Convolute* (see also *Involute*).—Usually employed to designate a wave or folds in opposite directions. A double involute.

157. *Conic Section.*—Having the form of or resembling a cone. Formed by cutting off a cone at any angle. See line A.

158. *Conoid.*—Anything that has a form resembling that of a cone.

159. *Cycloid.*—A curve, A, generated by a point, B, in the plane of a circle or wheel, C, when the wheel is rolled along a straight line.

160. *Ellipsoid.*—A solid, all plane sections of which are ellipses or circles.

161. *Epicycloid.*—A curve, A, traced by a point, B, in the circumference of a wheel, C, which rolls on the convex side of a fixed circle, D.

162. *Evolute.*—A curve, A, from which another curve, like B, on each of the inner ends of the lines C is made. D is a spool, and the lines C represent a thread at different positions. The thread has a marker, E, so that when the thread is wound on the spool the marker E makes the evolute line A.

163. *Focus.*—The center, A, of a circle; also one of the two centering points, B, of an ellipse or an oval.

164. *Gnome.* — The space included between the boundary lines of two similar parallelograms, the one within the other, with an angle in common.

165. *Hyperbola.* — A curve, A, formed by the section of a cone. If the cone is cut off vertically on the dotted line, A, the curve is a hyperbola. See *Parabola.*

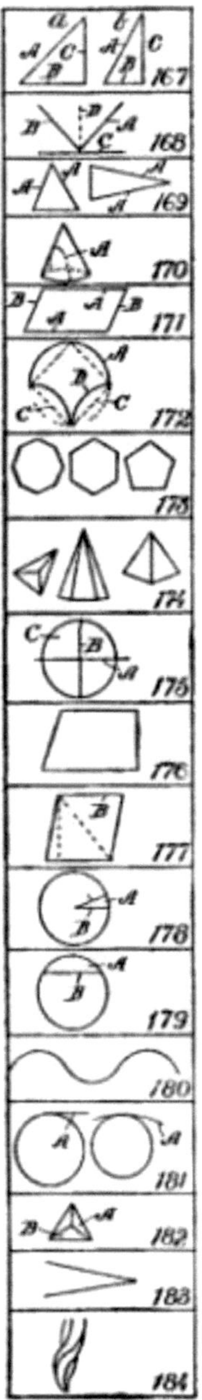

167. *Hypothenuse.* — The side, A, of a right-angled triangle which is opposite to the right angle B, C. A, regular triangle; C, irregular triangle.

[Pg 92] 168. *Incidence.* — The angle, A, which is the same angle as, for instance, a ray of light, B, which falls on a mirror, C. The line D is the perpendicular.

169. *Isosceles Triangle.* — Having two sides or legs, A, A, that are equal.

170. *Parabola.* — One of the conic sections formed by cutting of a cone so that the cut line, A, is not vertical. See *Hyperbola* where the cut line is vertical.

171. *Parallelogram.* — A right-lined quadrilateral figure, whose opposite sides, A, A, or B, B, are parallel and consequently equal.

172. *Pelecoid.* — A figure, somewhat hatchet-shaped, bounded by a semicircle, A, and two inverted quadrants, and equal to a square, C.

173. *Polygons.* — Many-sided and many with angles.

174. *Pyramid.* — A solid structure generally with a square base and having its sides meeting in an apex or peak. The peak is the vertex.

175. *Quadrant.* — The quarter of a circle or of the circumference of a circle. A horizontal line, A, and a vertical line, B, make the four quadrants, like C.

176. *Quadrilateral.* — A plane figure having four sides, and consequently four angles. Any figure formed by four lines.

177. *Rhomb.* — An equilateral parallelogram or a quadrilateral figure whose sides are equal and the opposite sides, B, B, parallel.

178. *Sector.* — A part, A, of a circle formed by two radial lines, B, B, and bounded at the end by a curve.

179. *Segment.* — A part, A, cut from a circle by a straight line, B. The straight line, B, is the chord or the *segmental line.*

180. *Sinusoid.* — A wave-like form. It may be regular or irregular.

181. *Tangent.* — A line, A, running out from the curve at right angles from a radial line.

182. *Tetrahedron.* — A solid figure enclosed or bounded by four triangles, like A or B. A plain pyramid is bounded by five triangles.

183. *Vertex.* — The meeting point, A, of two or more lines.

184. *Volute.* — A spiral scroll, used largely in architecture, which forms one of the chief features of the Ionic capital.

[Pg 93]

CHAPTER IX

MOLDINGS, WITH PRACTICAL ILLUSTRATIONS IN EMBELLISHING WORK

Moldings. — The use of moldings was early resorted to by the nations of antiquity, and we marvel to-day at many of the beautiful designs which the Phœnecians, the Greeks and the Romans produced. If you analyze the lines used you will be surprised to learn how few are the designs which go to make up the wonderful columns, spires, minarets and domes which are represented in the various types of architecture.

The Basis of Moldings. — Suppose we take the base type of moldings, and see how simple they are and then, by using these forms, try to build up or ornament some article of furniture, as an example of their utility.

The Simplest Molding. — In Fig. 185 we show a molding of the most elementary character known, being simply in the form of a band (A) placed below the cap. Such a molding gives to the article on which it is placed three distinct lines, C, D and E, If you stop to consider you will note that the molding, while it may add to the strength of the article, is primarily of service [Pg 94] because the lines and surfaces produce shadows, and therefore become valuable in an artistic sense.

The Astragal. — Fig. 186 shows the ankle-bone molding, technically called the *Astragal*. This form is round, and properly placed produces a good effect, as it throws the darkest shadow of any form of molding.

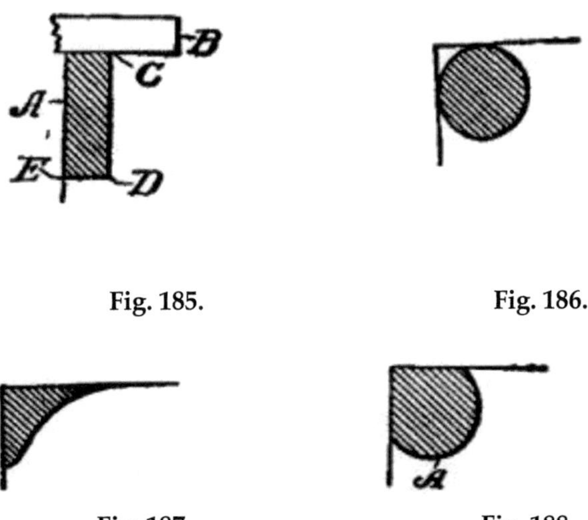

Fig. 185.

Fig. 186.

Fig. 187.

Fig. 188.

The Cavetto.—Fig. 187 is the cavetto, or round type. Its proper use gives a delicate outline, but it is principally applied with some other form of molding.

The Ovolo.—Fig. 188, called the ovolo, is a quarter round molding with the lobe (A) projecting downwardly. It is distinguished from [Pg 95] the astragal because it casts less of a shadow above and below.

The Torus.—Fig. 189, known as the torus, is a modified form of the ovolo, but the lobe (A) projects out horizontally instead of downwardly.

The Apophyges (Pronounced apof-i-ges).—Fig. 190 is also called the *scape*, and is a concaved type of molding, being a hollowed curvature used on columns where its form causes a merging of the shaft with the fillet.

Fig. 191. Cymatium.   Fig. 192. Ogee-Recta.

The Cymatium.—Fig. 191 is the cymatium (derived from the word cyme), meaning wave-like. This form must be in two curves, one inwardly and one outwardly.

The Ogee.—Fig. 192, called the ogee, is the most useful of all moldings, for two reasons: First, it may have the concaved surface uppermost, in which form it is called ogee recta—that is, right [Pg 96] side up; or it may be inverted, as in Fig. 193, with the concaved surface below, and is then called ogee reversa. Contrast these two views and you will note what a difference the mere inversion of the strip makes in the appearance. Second, because the ogee has in it, in a combined form, the outlines of nearly all the other types. The only advantage there is in using the other types is because you may thereby build up and space your work better than by using only one simple form.

Fig. 193. Ogee-Reversa.

Fig. 194. Bead or Reedy.

You will notice that the ogee is somewhat like the cymatium, the difference being that the concaved part is not so pronounced as in the ogee, and the convexed portion bulges much further than in the ogee. It is capable of use with other moldings, and may be reversed with just as good effect as the ogee.

[Pg 97] The Reedy.—Fig. 194 represents the reedy, or the bead—that is, it is made up of reeds. It is a type of molding which should not be used with any other pronounced type of molding.

The Casement (Fig. 195).—In this we have a form of molding used almost exclusively at the base of structures, such as columns, porticoes and like work.

Fig. 195. Casement.

Now, before proceeding to use these moldings, let us examine a Roman-Doric column, one of the most famous types of architecture produced. We shall see how the ancients combined moldings to produce grace, lights and shadows and artistic effects.

The Roman-Doric Column. — In Fig. 196 is shown a Roman-Doric column, in which the cymatium, the ovolo, cavetto, astragal and the ogee are used, together with the fillets, bases and caps, and it is interesting to study this because of its beautiful proportions.

[Pg 98]

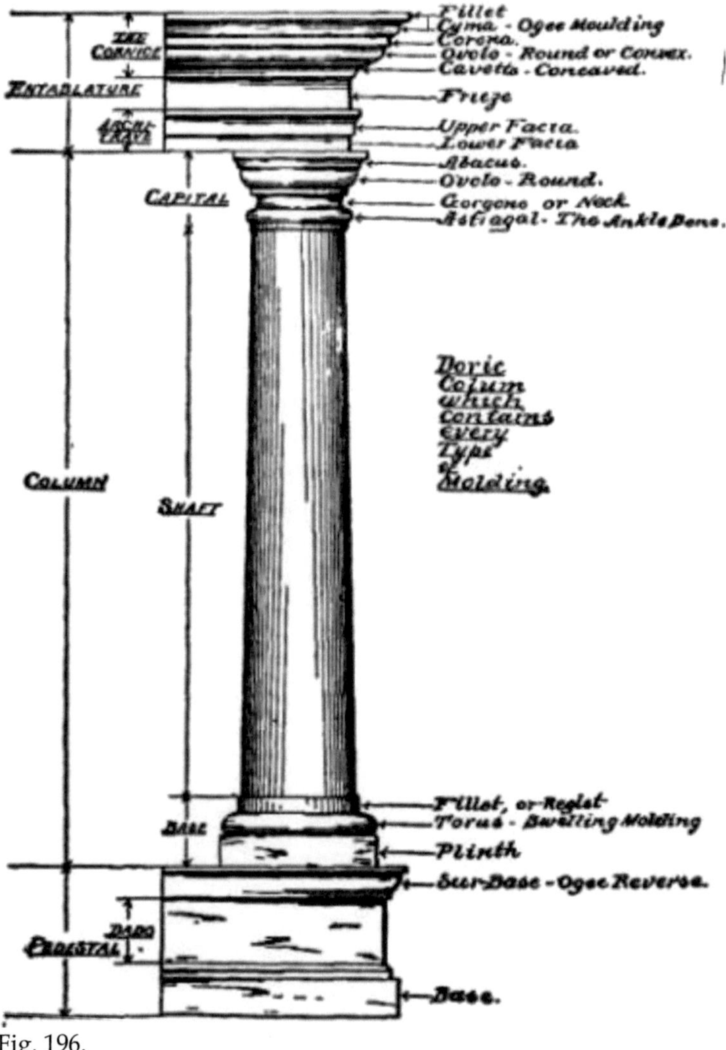

Fig. 196.

The pedestal and base are equal in vertical dimensions to the entablature and capital. The entablature is but slightly narrower than the pedestal; [Pg 99] and the length of the column is, approximately, four times the height of the pedestal. The base of the shaft, while

112

larger diametrically than the capital, is really shorter measured vertically. There is a reason for this. The eye must travel a greater distance to reach the upper end of the shaft, and is also at a greater angle to that part of the shaft, hence it appears shorter, while it is in reality longer. For this reason a capital must be longer or taller than the base of a shaft, and it is also smaller in diameter.

It will be well to study the column not only on account of the wonderful blending of the various forms of moldings, but because it will impress you with a sense of proportions, and give you an idea of how simple lines may be employed to great advantage in all your work.

Lessons from the Doric Column.—As an example, suppose we take a plain cabinet, and endeavor to embellish it with the types of molding described, and you will see to what elaboration the operation may be carried.

Applying Molding.—Let Fig. 197 represent the front, top and bottom of our cabinet; and the first thing we shall do is to add a base (A) and a cap (B). Now, commencing at the top, suppose we utilize the simplest form of molding, the band.

This we may make of any desired width, as [Pg 100] shown in Fig. 198. On this band we can apply the ogee type (Fig. 199) right side up.

But for variation we may decide to use the ogee reversed, as in Fig. 200. This will afford us something else to think about and will call upon our powers of initiative in order to finish off the lower margin or edge of the ogee reversa.

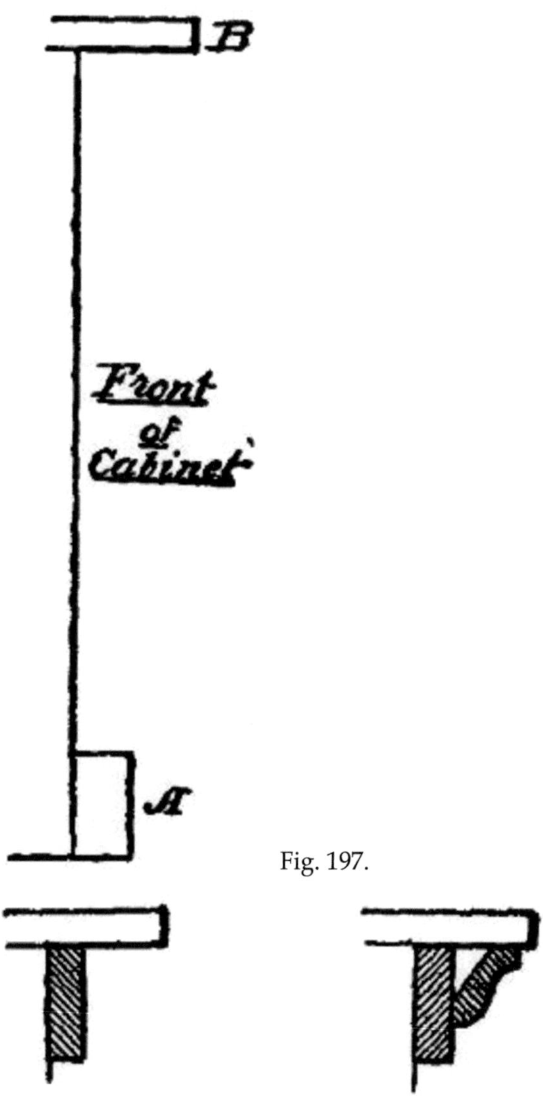

Fig. 197.

If we take the ogee recta, as shown in Fig. 201, we may use the cavetto, or the ovolo (Fig. 202); but if we use the ogee reversa we must use a convex molding like the cavetto at one base, and [Pg 101] a convex molding, like the torus or the ovolo, at the other base.

In the latter (Fig. 202) four different moldings are used with the ogee as the principal structure.

Base Embellishments.—In like manner (Fig. 204) the base may have the casement type first attached in the corner, and then the ovolo, or the astragal added, as in Fig. 203.

Straight-faced Moldings.—Now let us carry the principle still further, and, instead of using various type of moldings, we will employ nothing but straight strips of wood. This treatment will soon indicate to you that the true mechanic or artisan is he who can take advantage of whatever he finds at hand.

Let us take the same cabinet front (Fig. 205), and below the cap (A) place a narrow strip (B), the lower corner of which has been chamfered off, as at C. Below the strip B is a thinner strip (D), vertically disposed, and about two-thirds its width. The lower corner of this is also chamfered, as at [Pg 102] F. To finish, apply a small strip (G) in the corner, and you have an embellished top that has the appearance, from a short distance, of being made up of molding.

Plain Molded Base.—The base may be treated in the same manner. The main strip (4) has its upper corner chamfered off, as at I, and on this is nailed a thin, narrow finishing strip (J). The upper part or molded top, in this case, has eleven distinct lines, and the base has six lines. By experimenting you may soon put together the most available kinds of molding strips.

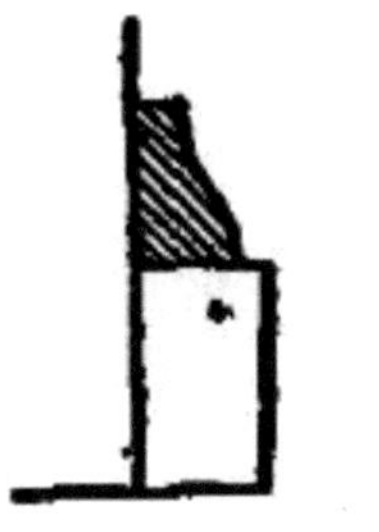 

Fig. 203.   Fig. 204.

Diversified Uses. — For a great overhang you may use the cavetto, or the apophyges, and below that the astragal or the torus; and for the base the casement is the most serviceable molding, and it may be finished off with the ovolo or the cymatium.

Pages of examples might be cited to show the variety and the diversification available with different types.

[Pg 103] Shadows Cast by Moldings. — Always bear in mind that a curved surface makes a blended shadow. A straight, flat or plain surface does not, and it is for that reason the concaved and the convexed surfaces, brought out by moldings, become so important.

Fig. 205.

A little study and experimenting will soon teach you how a convex, a concave or a flat surface, and a corner or corners should be arranged relatively to each other; how much one should project beyond the other; and what the proportional widths of the different molding bands should be. An entire volume would scarcely exhaust this subject.

[Pg 104]

CHAPTER X

AN ANALYSIS OF TENONING, MORTISING, RABBETING AND BEADING

In the chapter on How Work is Laid Out, an example was given of the particular manner pursued in laying out mortises and tenons, and also dovetailed work. I deem it advisable to add some details to the subject, as well as to direct attention to some features which do not properly belong to the laying out of work.

Where Mortises Should Be Used.—Most important of all is a general idea of places and conditions under which mortises should be resorted to. There are four ways in which different members may be secured to each other. First, by mortises and tenons; second, by a lap-and-butt; third, by scarfing; and, fourth, by tonguing and grooving.

Depth of Mortises.—When a certain article is to be made, the first consideration is, how the joint or joints shall be made. The general rule for using the tenon and mortise is where two parts are joined wherein the grains of the two [Pg 105] members run at right angles to each other, as in the following figure.

Rule for Mortises.—Fig. 206 shows such an example. You will notice this in doors particularly, as an example of work.

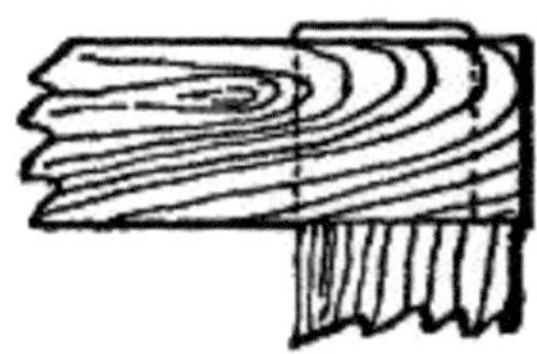

Fig. 206.

Fig. 207.

The next consideration is, shall the mortises be cut entirely through the piece? This is answered by the query as to whether or not the end of the tenon will be exposed; and usually, if a smooth finish is required, the mortise should not go through the member. In a door, however, the tenons are exposed at the edges of the door, and are, therefore, seen, so that we must apply some other rule. The one universally adopted is, that where, as in a door stile, it is broad and comparatively thin, or where the member having the mortise [Pg 106] in its edge is much thinner than its width, the mortise should go through from edge to edge.

The reason for this lies in the inability to sink the mortises through the stile (A, Fig. 207) perfectly true, and usually the job is turned out something like the illustration shows. The side of the rail (B) must be straight with the side of the stile. If the work is done by machinery it results in accuracy unattainable in hand work.

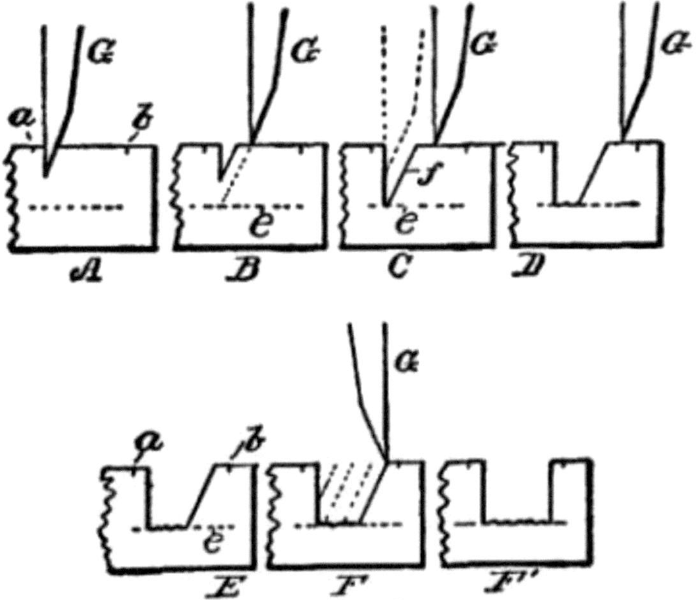

Fig. 208.

True Mortise Work.—The essense of good joining work is the ability to sink the chisel true with the side of the member. More uneven work is produced by haste than by inability. The tendency [Pg 107] of all beginners is to strike the chisel too hard, in order the more quickly to get down to the bottom of the mortise. Hence, bad work follows.

Steps in Cutting Mortises.—Examine Fig. 208, which, for convenience, gives six successive steps in making the mortise. The marks a, b designate the limits, or the length, of the mortise. The chisel (C) is not started at the marking line (A), but at least an eighth of an inch from it. The first cut, as at B, gives a starting point for the next cut or placement of the chisel. When the second cut (B) has thus been made, the chisel should be turned around, as in dotted line d, position C, thereby making a finish cut down to the bottom of the mortise, line e, so that when the fourth cut has been made along line f, we are ready for the fifth cut, position C; then the sixth cut, position D, which leaves the mortise as shown at E. Then turn the chisel to the position shown at F, and cut down the last end of the mortise square, as shown in G, and clean out the mortise well before making the finishing cuts on the marking lines (*a*, *b*). The particular reason for cleaning out the mortise before making the finish cuts is, that the corners of the mortise are used as fulcrums for the chisels, and the eighth of an inch stock still remaining protects the corners.

[Pg 108] Things to Avoid in Mortising. — You must be careful to refrain from undercutting as your chisel goes down at the lines *a*, *b*, because if you commit this error you will make a bad joint.

As much care should be exercised in producing the tenon, although the most common error is apt to occur in making the shoulder. This should be a trifle undercut.

Fig. 209.

See the lines (A, Fig. 209), which illustrate this.

Lap-and-Butt Joint. — The lap-and-butt is the form of uniting members which is most generally used to splice together timbers, where they join each other end to end.

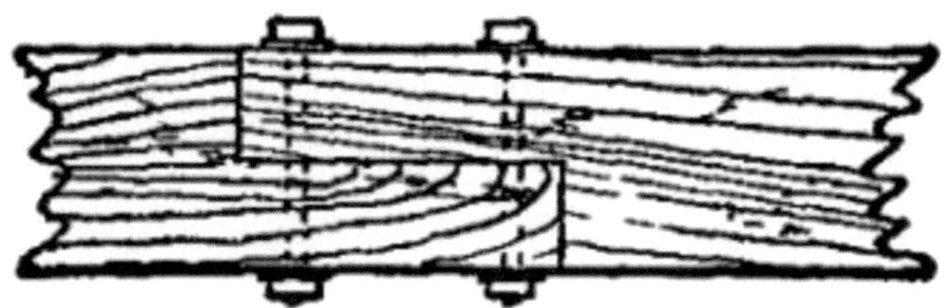

Fig. 210.

Bolts are used to secure the laps.

But the lap-and-butt form is also used in doors and in other cabinet work. It is of great service in paneling.

[Pg 109] A rabbet is formed to receive the edge of the panel, and a molding is then secured to the other side on the panel, to hold the latter in place.

Scarfing. — This method of securing members together is the most rigid, and when properly performed makes the joint the strongest part of the timber. Each member (A, Fig. 212) has a step diagonally cut (B), the two steps being on different planes, so they form a hook joint, as at C, and as each point or terminal has a blunt end, the members are so constructed as to withstand a longitudinal strain in either direction. The overlapping plates (D) and the bolts (E) hold the joint rigidly.

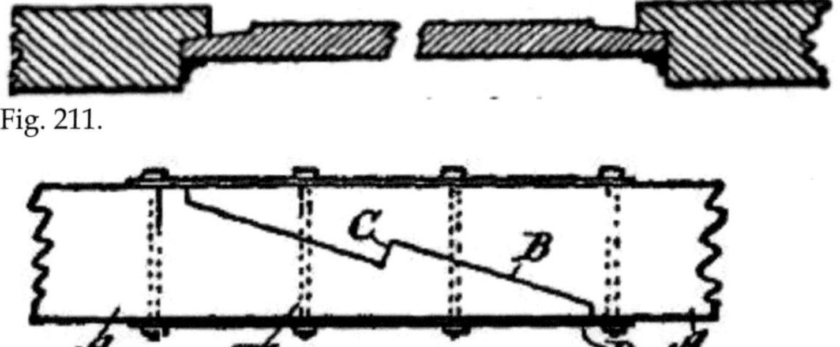

Fig. 211.

Fig. 212.

The Tongue and Groove. — This form of uniting members has only a limited application. It is [Pg 110] serviceable for floors, table tops, paneling, etc. In Fig. 213, a door panel is shown, and the door mullions (B) are also so secured to the rail (C). The tongue-and-groove method is never used by itself. It must always have some support or reinforcing means.

Fig. 213.

Fig. 214.   Fig. 215.

Beading.—This part of the work pertains to surface finishings, and may or may not be used in connection with rabbeting.

Figs. 214 and 215 show the simplest and most generally adopted forms in which it is made and used in connection with rabbeting, or with the tongue and groove. The bead is placed on one or both sides of that margin of the board (Fig. 214) which has the tongue, and the adjoining board has the usual flooring groove to butt against and receive the tongue. It is frequently the case that a blind bead, as in Fig. 215, runs through [Pg 111] the middle of the board, so as to give the appearance of narrow strips when used for wainscoting, or for ceilings. The beads also serve to hide the joints of the boards.

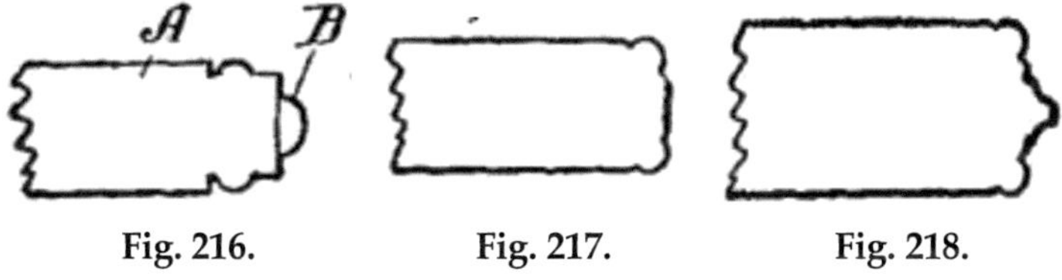

Fig. 216.   Fig. 217.   Fig. 218.

Ornamental Bead Finish.—These figures show how the bead may be used for finishing corners, edges and projections. Fig. 216 has a bead at each corner of a stile (A), and a finishing strip of half-round material (B) is nailed to the flat edge. Fig. 217 has simply the corners themselves beaded, and it makes a most serviceable finish for the edges of projecting members.

Fig. 218, used for wider members, has the corners beaded and a fancy molding (C); or the reduced edge of the stile itself is rounded off.

Fig. 219.   **Fig. 220.**

The Bead and Rabbet.—A more amplified form of work is available where the rabbet plane is used with the beader. These two planes together [Pg 112] will, if properly used, offer a strong substitute for molding and molding effects.

Fig. 219 has both sides first rabbeted, as at A, and the corners then beaded, as at B, with the reduced part of the member rounded off, as at C. Or, as in Fig. 220, the reduced edge of the member may have the corners beaded, as at D, and the rabbeted corners filled in with a round or concaved moulding (E).

Shading with Beads and Rabbets.—You will see from the foregoing, that these embellishments are serviceable because they provide the article with a large number of angles and surfaces to cast lights and shadows; and for this reason the boy should strive to produce the effects which this class of work requires.

[Pg 113]

CHAPTER XI

HOUSE BUILDING

House building is the carpenter's craft; cabinet-making the joiner's trade, yet both are so intimately associated, that it is difficult to draw a line. The same tools, the same methods and the same materials are employed.

There is no trade more ennobling than home building. It is a vocation which touches every man and woman, and to make it really an art is, or should be, the true aspiration of every craftsman.

The House and Embellishments.—The refined arts, such as sculpture and painting, merely embellish the home or the castle, so that when we build the structure it should be made with an eye not only to comfort and convenience, but fitting in an artistic and æsthetic sense. It is just as easy to build a beautiful home as an ugly, ungainly, illy proportioned structure.

Beauty Not Ornamentation.—The boy, in his early training, should learn this fundamental truth, that beauty, architecturally, does not depend upon ornamentation. Some of the most beautiful structures in the world are very plain. [Pg 114] Beauty consists in proportions, in proper correlation of parts, and in adaptation for the uses to which the structure is to be put.

Plain Structures.—A house with a plain façade, having a roof properly pitched and with a simple cornice, if joined to a wing which is not ungainly or out of proper proportions, is infinitely more beautiful than a rambling structure, in which one part suggests one order of architecture and the other part some other type or no type at all, and in which the embellishments are out of keeping with the size or pretensions of the house.

Colonial Type.—For real beauty, on a larger scale, there is nothing to-day which equals the old Colonial type with the Corinthian columns and entablature. The Lee mansion, now the National Cemetery, at Washington, is a fine example. Such houses are usually square or rectangular in plan, severely plain, with the whole ornamentation consisting of the columns and the portico. This type presents an appearance of massiveness and grandeur and is an excellent illustration of a form wherein the main characteristic of the structure is concentrated or massed at one point.

The Church of the Madelaine, Paris, is another striking example of this period of architecture.

Of course, it would be out of place with cottages and small houses, but it is well to study and to [Pg 115] know what forms are most available and desirable to adopt, and particularly to know something of the art in which you are interested.

The Roof the Keynote.—Now, there is one thing which should, and does, distinguish the residence from other types of buildings, excepting churches. It is the roof. A house is dominated by its covering. I refer to the modern home. It is not true with the Colonial or the Grecian types. In those the façade or the columns and cornices predominate over everything else.

Bungalow Types.—If you will take up any book on bungalow work and note the outlines of the views you will see that the roof forms the main element or theme. In fact, in most buildings of this kind everything is submerged but the roof and roof details. They are made exceedingly flat, with different pitches with dormers and gables intermingled and indiscriminately placed, with cornices illy assorted and of different kinds, so that the multiplicity of diversified details gives an appearance of great elaboration. Many of those designs are monstrosities and should, if possible, be legally prohibited.

I cannot attempt to give even so much as an outline of what constitutes art in its relation to building, but my object is to call attention to this phase of the question, and as you proceed in [Pg 116] your studies and your work you will realize the value and truthfulness of the foregoing observations.

General House Building.—We are to treat, generally, on the subject of house building, how the work is laid out, and how built, and in doing so I shall take a concrete example of the work. This can be made more effectual for the purpose if it is on simple lines.

Building Plans.—We must first have a plan; and the real carpenter must have the ability to plan as well as to do the work. We want a five-room house, comprising a parlor, dining room, two bedrooms, a kitchen and a bathroom. Just a modest little home, to which we can devote our spare hours, and which will be neat and comfortable when finished. It must be a one-story house, and that fact at once settles the roof question. We can make the house perfectly square in plan, or rectangular, and divide up the space into the proper divisions.

The Plain Square Floor Plan will first be taken up, as it is such an easy roof to build. Of course, it is severely plain.

Fig. 221 shows our proposed plan, drawn in the rough, without any attempts to measure the different apartments, and with the floor plan exactly square. Supposing we run a hall (A) through the middle. On one side of this let us plan for a dining room and a kitchen, a portion of the kitchen space to be given over to a closet and a bathroom.

[Pg 117]

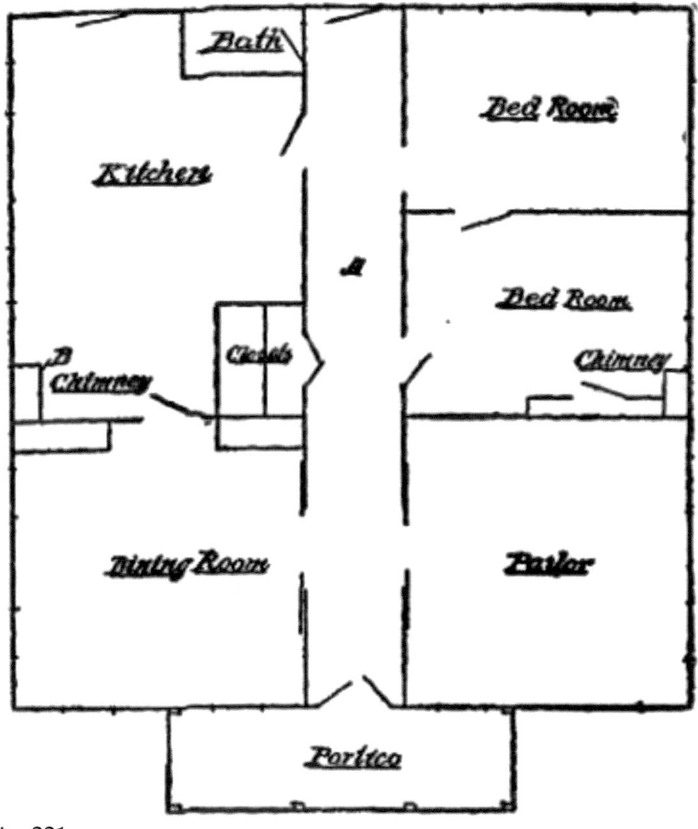

Fig. 221.

The chimney (B) must be made accessible from both rooms. On the other side of the hallway the space is divided into a parlor and two bedrooms.

[Pg 118] The Rectangular Plan.—In the rectangular floor plan (Fig. 222) a portion of the floor space is cut out for a porch (A), so that we may use the end or the side for the entrance. Supposing we use the end of the house for this purpose. The entrance room (B) may be a bedroom, or a reception and living room, and to the rear of this room is the dining room, connected with the reception room by a hall (C). This hall also leads to the kitchen and to the bathroom, as well as to the other bedroom. The parlor is connected with the entrance room (B), and also with the bedroom. All of this is optional, of course.

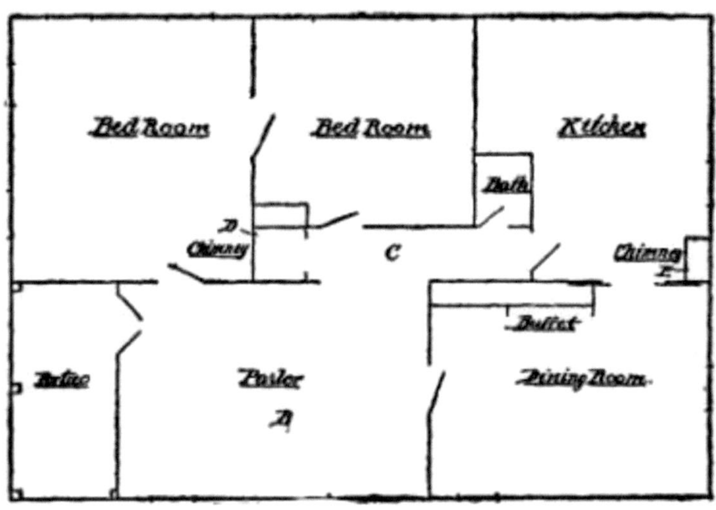

Fig. 222.

There are also two chimneys, one chimney (D) having two flues and the other chimney (E) having three flues, so that every room is accommodated.

[Pg 119]

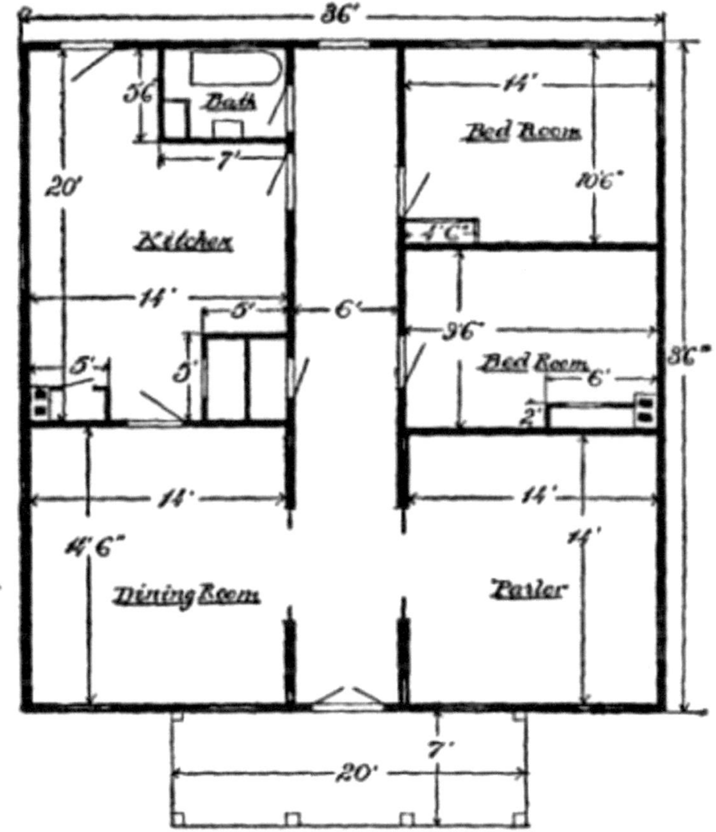

Fig. 223.

Room Measurements.—We must now determine the dimensions of each room, and then how we shall build the roof.

In Figs. 223 and 224, we have now drawn out [Pg 120] in detail the sizes, the locations of the door and windows, the chimneys and the closets, as well as the bathroom. All this work may be changed or modified to suit conditions and the taste of the designer.

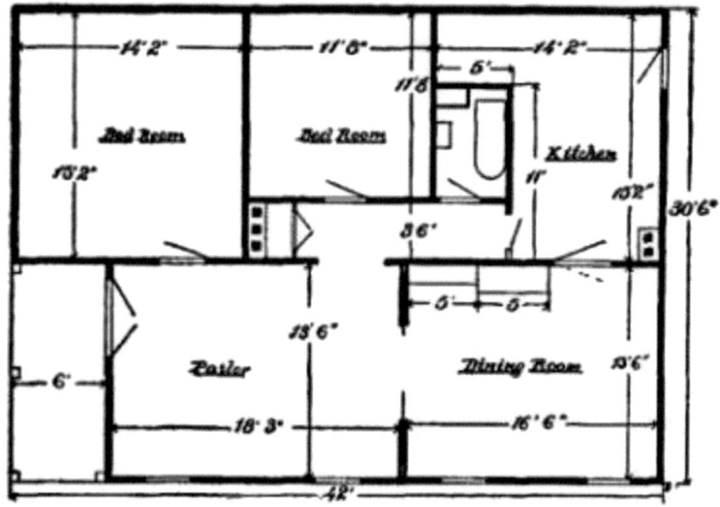

Fig. 224.

Front and Side Lines.—From the floor diagram, and the door and window spaces, as marked out, we may now proceed to lay out rough front and side outlines of the building. The ceilings are to be 9 feet, and if we put a rather low-pitched roof on the square structure (Fig. 223) the front may look something like Fig. 225, and a greater pitch given to the rectangular plan (Fig. 224) will present a view as shown in Fig. 226.

[Pg 121]

Fig. 225.

Fig. 226.

The Roof.—The pitch of the roof (Fig. 225) is what is called "third pitch," and the roof (Fig. 226) has a half pitch. A "third" pitch is determined as follows:

[Pg 122] Roof Pitch.—In Fig. 227 draw a vertical line (A) and join it by a horizontal line (B). Then strike a circle (C) and step it off into three parts. The line (D), which intersects the first mark (E) and the angle of the lines (A, B), is the pitch.

131

In Fig. 228 the line A is struck at 15 degrees, which is halfway between lines B and C, and it is, therefore, termed "half-pitch."

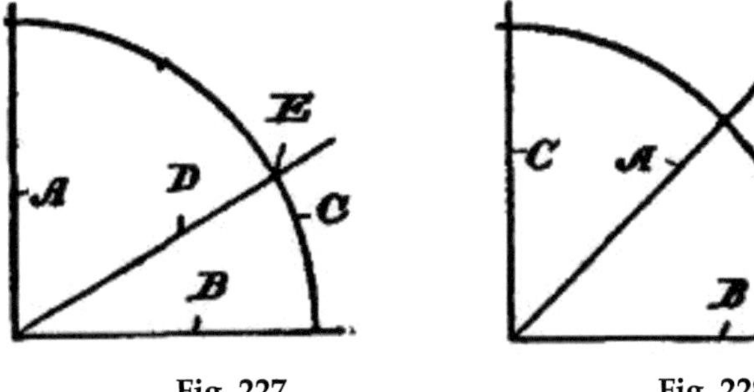

Fig. 227.   Fig. 228.

Thus, we have made the ground plans, the elevations and the roofs as simple as possible. Let us proceed next with the details of the building.

The Foundation.—This may be of brick, stone or concrete, and its dimensions should be at least 1½ inches further out than the sill.

The Sills.—We are going to build what is called a "balloon frame"; and, first, we put down the sills, which will be a course of 2" × 6", or 2" × 8" joists, as in Fig. 229.

The Flooring Joist.—The flooring joists (A) are then put down (Fig. 230). These should extend [Pg 123] clear across the house from side to side, if possible, or, if the plan is too wide, they should be lapped at the middle wall and spiked together. The ends should extend out flush with the outer margins of the sills, as shown, but in putting down the first and last sill, space must be left along the sides of the joist of sufficient width to place the studding.

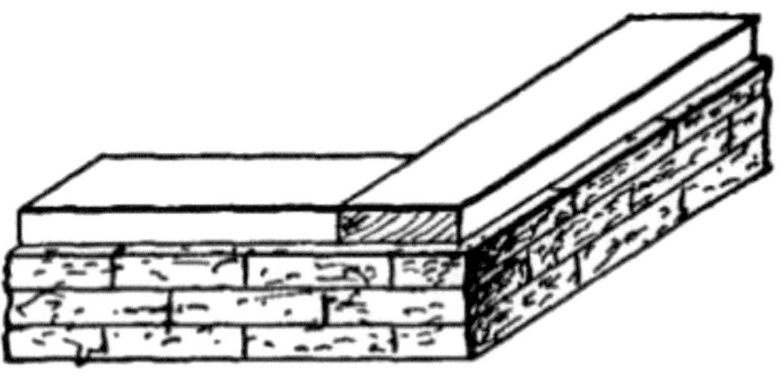

Fig. 229.

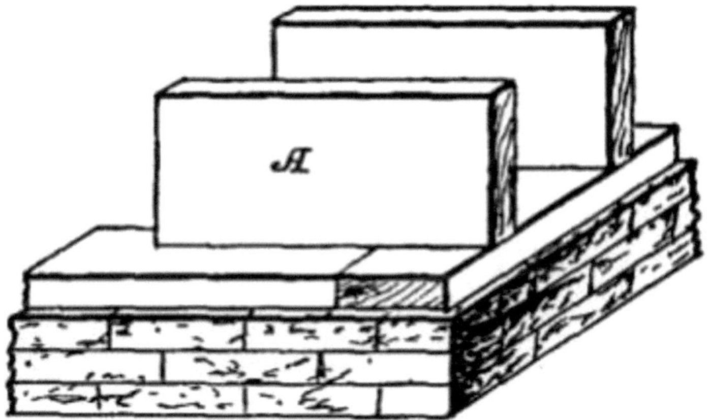

Fig. 230.

The Studding. — The next step is to put the studding into position. 4" × 4" must be used for corners and at the sides of door and window openings. [Pg 124] 4" × 6" may be used at corners, if preferred. Consult your plan and see where the openings are for doors and windows. Measure the widths of the door and window frames, and make a measuring stick for this purpose. You must leave at least one-half inch clearance for the window or door frame, so as to give sufficient room to plumb and set the frame.

Setting Up. — First set up the corner posts, plumbing and bracing them. Cut a top plate for each side you are working on.

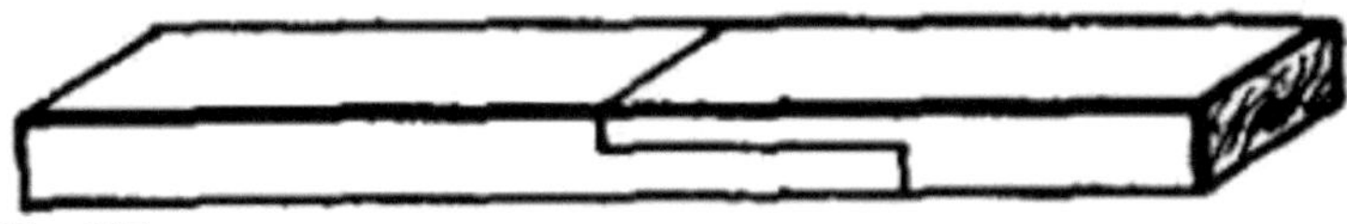

Fig. 231.

The Plate. — As it will be necessary in our job to use two or more lengths of 2" × 4" scantling for the plate, it will be necessary to join them together. Do this with a lap-and-butt joint (Fig. 231).

Then set up the 4" × 4" posts for the sides of the doors and windows, and for the partition walls.

The plate should be laid down on the sill, and marked with a pencil for every scantling to correspond with the sill markings. The plate is then put on and spiked to the 4" × 4" posts.

Intermediate Studding. — It will then be an [Pg 125] easy matter to put in the intermediate 2" × 4" studding, placing them as nearly as possible 16 inches apart to accommodate the 48-inch plastering lath.

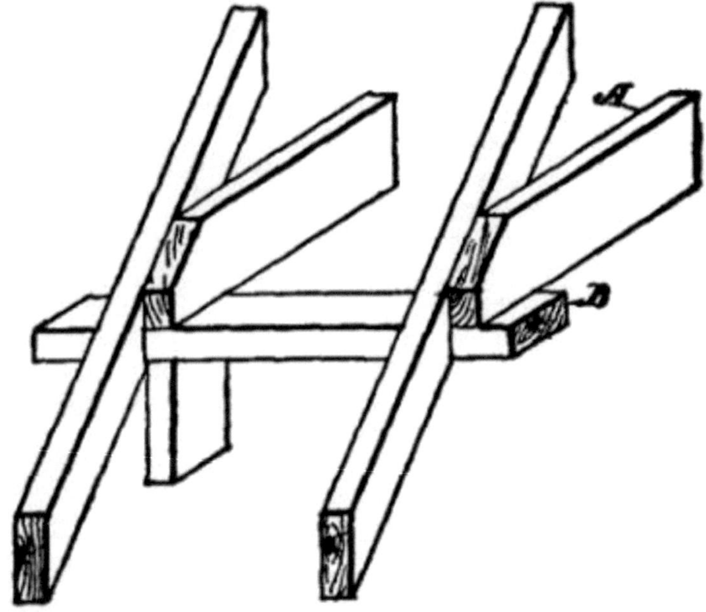

Fig. 232.

Wall Headers.—When all the studding are in you will need headers above and rails below the windows and headers above all the doors, so that you will have timbers to nail the siding to, as well as for the lathing.

Ceiling Joists.—We are now ready for the ceiling joists, which are, usually, 2" × 6", unless there is an upper floor. These are laid 16 inches apart from center to center, preferably parallel with the floor joist.

It should be borne in mind that the ceiling [Pg 126] joist must always be put on with reference to the roof.

Thus, in Fig. 232, the ceiling joists (A) have their ends resting on the plate (B), so that the rafters are in line with the joists.

Braces.—It would also be well, in putting up the studding, to use plenty of braces, although for a one-story building this is not so essential as in two-story structures, because the weather boarding serves as a system of bracing.

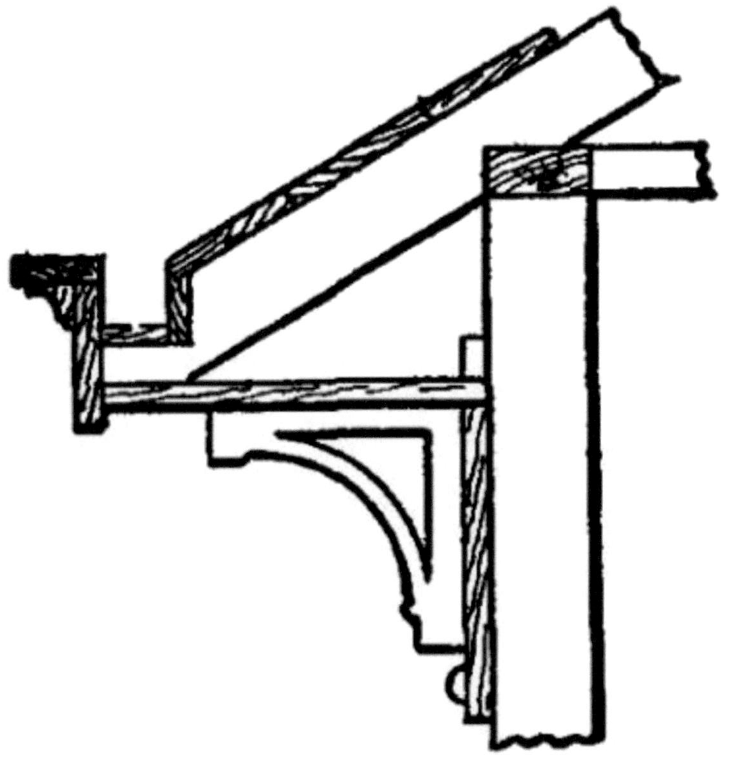

Fig. 233.

The Rafters. — These may be made to provide for the gutter or not, as may be desired. They should be of 2" × 4" scantling.

The Gutter. — In Fig. 233 I show a most serviceable way to provide for the gutter. A V-shaped notch is cut out of the upper side of the rafter, in which is placed the floor and a side. This [Pg 127] floor piece is raised at one end to provide an incline for the water.

A face-board is then applied and nailed to the ends of the rafters. This face-board is surmounted by a cap, which has an overhang, beneath which is a molding of any convenient pattern. The face-board projects down at least two inches below the angled cut of the rafter, so that when the base-board is applied, the lower margin of the face-board will project one inch below the base.

Fig. 234.

This base-board is horizontal, as you will see. The facia-board may be of any desired width, and a corner molding should be added. It is optional to use the brackets, but if added they should be spaced apart a distance not greater than twice the height of the bracket.

A much simpler form of gutter is shown in Fig. 234, in which a V-shaped notch is also cut in the [Pg 128] rafter, and the channel is made by the pieces. The end of the rafter is cut at right angles, so the face-board is at an angle. This is also surmounted by an overhanging cap and a molding. The base is nailed to the lower edges of the rafters, and the facia is then applied.

Fig. 234a.

In Fig. 234*a* the roof has no gutter, so that the end of the rafter is cut off at an angle and a molding applied on the face-board. The base is nailed to the rafters. This is the cheapest and simplest form of structure for the roof.

Setting Door and Window Frames. — The next step in order is to set the door and window frames preparatory to applying the weather boarding. It is then ready for the roof, which should be put on before the floor is laid.

Plastering and Inside Finish. — Next in order is the plastering, then the base-boards and the [Pg 129] casing; and, finally, the door and windows should be fitted into position.

138

Enough has been said here merely to give a general outline, with some details, how to proceed with the work.

[Pg 130]

CHAPTER XII

BRIDGES, TRUSSED WORK AND LIKE STRUCTURES

Bridges.—Bridge building is not, strictly, a part of the carpenter's education at the present day, because most structures of this kind are now built of steel; but there are certain principles involved in bridge construction which the carpenter should master.

Self-supporting Roofs.—In putting up, for instance, self-supporting roofs, or ceilings with wide spans, and steeples or towers, the bridge principle of trussed members should be understood.

The most simple bridge or trussed form is the well-known A-shaped arch.

Fig. 235.

Common Trusses.—One form is shown in Fig. 235, with a vertical king post. In Fig. 236 there are two vertical supporting members, called queen posts, used in longer structures. Both of these [Pg 131] forms are equally well adapted for small bridges or for roof supports.

The Vertical Upright Truss.—This form of truss naturally develops into a type of wooden bridge known all over the country, as its framing is simple, and calculations as to its capacity to sustain loads may readily be made. Figs. 237, 238 and 239 illustrate these forms.

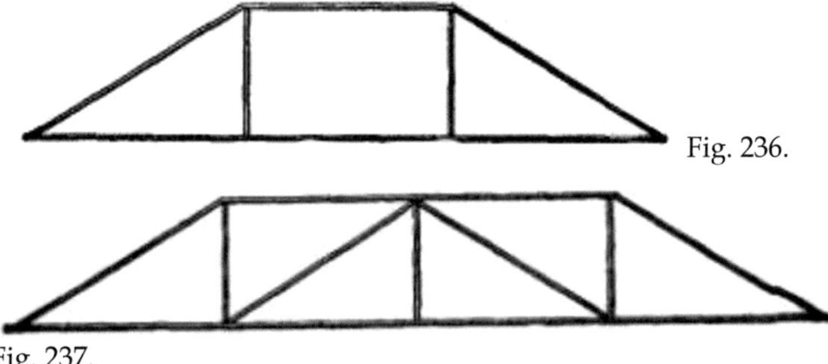

Fig. 236.

Fig. 237.

The Warren Girder.—Out of this simple truss grew the Warren girder, a type of bridge particularly adapted for iron and steel construction.

This is the simplest form for metal bridge truss, or girder. It is now also largely used in steel buildings and for other work requiring strength with small weight.

[Pg 132]

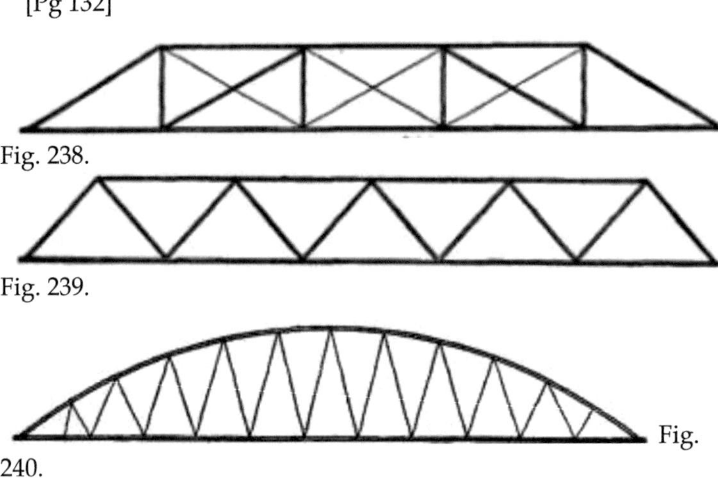

Fig. 238.

Fig. 239.

Fig. 240.

The Bowstring Girder.—Only one other form of [Pg 133] bridge truss need be mentioned here, and that is the *bowstring* shown in Fig. 240.

140

In this type the bow receives the entire compression thrust, and the chords act merely as suspending members.

Fundamental Truss Form.—In every form of truss, whether for building or for bridge work, the principles of the famous A-truss must be employed in some form or other; and the boy who is experimentally inclined will readily evolve means to determine what degree of strength the upper and the lower members must have for a given length of truss to sustain a specified weight.

There are rules for all these problems, some of them very intricate, but all of them intensely interesting. It will be a valuable addition to your knowledge to give this subject earnest study.

[Pg 134]

CHAPTER XIII

THE BEST WOODS FOR THE BEGINNER

In this place consideration will be given to some of the features relating to the materials to be employed, particularly with reference to the manner in which they can be worked to the best advantage, rather than to their uses.

The Best Woods.—The prime wood, and the one with which most boys are familiar, is white pine. It has an even texture throughout, is generally straight grained, and is soft and easily worked. White pine is a wood requiring a very sharp tool. It is, therefore, the best material for the beginner, as it will at the outset teach him the important lesson of keeping the tools in a good, sharp condition.

Soft Woods.—It is also well for the novice to do his initial work with a soft wood, because in joining the parts together inaccuracies may be easily corrected. If, for instance, in mortising and tenoning, the edge of the mortised member is not true, or, rather, is not "square," the shoulder of the tenon on one side will abut before the other side does, and thus leave a crack, if the wood is hard. If the wood is soft there is always enough yield to [Pg 135] enable the workman to spring it together. Therefore, until you have learned how to make a true joint, use soft wood.

Poplar is another good wood for the beginner, as well as redwood, a western product.

Hard Woods.—Of the hard woods, cherry is the most desirable for the carpenter's tool. For working purposes it has all the advantages of a soft wood, and none of its disadvantages. It is not apt to warp, like poplar or birch, and its shrinking unit is less than that of any other wood, excepting redwood. There is practically no shrinkage in redwood.

The Most Difficult Woods.—Ash is by far the most difficult wood to work. While not as hard as oak, it has the disadvantage that the entire board is seamed with growth ribs which are extremely hard, while the intervening layers between these ribs are soft, and have open pores, so that, for instance, in making a mortise, the chisel is liable to follow the hard ribs, if the grain runs at an angle to the course of the mortise.

The Hard-ribbed Grain in Wood.—This peculiarity of the grain in ash makes it a beautiful wood when finished. Of the light-colored woods, oak only excels it, because in this latter wood each year's growth shows a wider band, and the interstices between the ribs have stronger contrasting [Pg 136] colors than ash; so that in filling the surface, before finishing it, the grain of the wood is brought out with most effective clearness and with a beautifully blended contrast.

The Easiest Working Woods.—The same thing may be said, relatively, concerning cherry and walnut. While cherry has a beautiful finishing surface, the blending contrasts of colors are not so effective as in walnut.

Oregon pine is extremely hard to work, owing to the same difficulties experienced in handling ash; but the finished Oregon pine surface makes it a most desirable material for certain articles of furniture.

Do not attempt to employ this nor ash until you have mastered the trade. Confine yourself to pine, poplar, cherry and walnut.

These woods are all easily obtainable everywhere, and from them you can make a most creditable variety of useful articles.

Sugar and maple are two hard woods which may be added to the list. Sugar, particularly, is a good-working wood, but maple is more difficult. Spruce, on the other hand, is the strongest and toughest wood, considering its weight, which is but a little more than that of pine.

Differences in the Working of Woods.—Different woods are not worked with equal facility by [Pg 137] all the tools. Oak is an easy wood to handle with a saw, but is, probably, aside from ash, the most difficult wood known to plane.

Ash is hard for the saw or the plane. On the other hand, there is no wood so easy to manipulate with the saw or plane as cherry. Pine is easily worked with a plane, but difficult to saw; not on account of hardness, but because it is so soft that the saw is liable to tear it.

Forcing Saws in Wood.—One of the reasons why the forcing of saws is such a bad practice will be observed in cutting white or yellow pine. For cross-cutting, the saw should have fine teeth, not heavily set, and evenly filed. To do a good job of cross-cutting, the saw must be held at a greater angle, or should lay down flatter than in ripping, as by so doing the lower side of the board will not break away as much as if the saw should be held more nearly vertical.

These general observations are made in the hope that they will serve as a guide to enable you to select your lumber with some degree of intelligence before you commence work.

[Pg 138]

CHAPTER XIV

WOOD TURNING

Advantages of Wood Turning.—This is not, strictly, in the carpenter's domain; but a knowledge of its use will be of great service in the trade, and particularly in cabinet making. I urge the ingenious youth to rig up a wood-turning lathe, for the reason that it is a tool easily made and one which may be readily turned by foot, if other power is not available.

Simple Turning Lathe.—A very simple turning lathe may be made by following these instructions:

The Rails.—Procure two straight 2" × 4" scantling (A), four feet long, and planed on all sides. Bore four ⅜-inch holes at each end, as shown, and 10 inches from one end four more holes. A plan of these holes is shown in B, where the exact spacing is indicated. Then prepare two pieces 2" × 4" scantling (C), planed, 42 inches long, one end of each being chamfered off, as at 2, and provided with four bolt holes. Ten inches down, and on the same side, with the chamfer (2) is a cross gain (3), the same angle as the chamfer. Midway between the cross gain (3) and the lower end of the leg is [Pg 139] a gain (4) in the edge, at right angles to the cross gain (3).

The Legs.—Now prepare two legs (D) for the tail end of the frame, each 32 inches long, with a chamfer (5) at one end, and provided with four bolt holes. At the lower end bore a bolt hole for the cross base piece. This piece (E) is 4" × 4", 21 inches long, and has a bolt hole at each end and one near the middle. The next piece (F) is 2" × 4", 14½ inches long, provided with a rebate (6) at each end, to fit the cross gains (4) of the legs (C). Near the middle is a journal block (7).

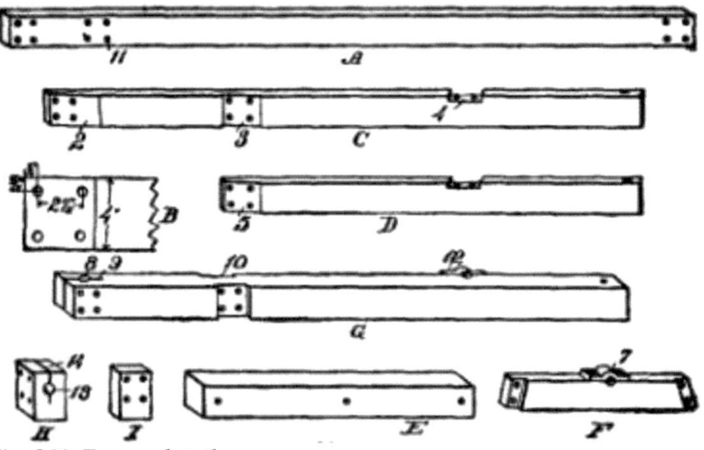

Fig. 241. Frame details.

Centering Blocks.—Next provide a 4" × 4" piece (G), 40 inches long, through which bore a ¾-inch hole (8), 2 inches from the upper end, and [Pg 140] four bolt holes at right angles to the shaft hole (8). Then, with a saw split down this bearing, as shown at 9, to a point 4 inches from the end. Ten inches below the upper end prepare two cross gains (10), each an inch deep and four inches wide. In these gains are placed the top rails (A), so the bolt holes in the gains (10) will coincide with the bolt holes (11) in the piece A. Below the gains (10) this post has a journal block (12), intended to be in line with the journal block (7) of the piece F.

Fig. 242. Tail Stock.

Then make a block (H) 2" × 4", and 6 inches long. This also must have a shaft hole (B), and a saw kerf (14), similar to the arrangement on the upper end of the post (G); also bore four bolt holes, as shown. This block rests between the upper ends of the lugs (C).

Another block (I), 2" × 4", and 6 feet long, with four bolt holes, will be required for the tail end of the frame, to keep the rails (A) two inches apart at that end.

The Tail Stock. — This part of the structure is made of the following described material:

[Pg 141] Procure a scantling (J), planed, 4" × 4", 24 inches long, the upper end of which is to be provided with four bolt holes, and a centering hole (15). At the lower end of the piece is a slot (16) 8 inches long and 1½ inches wide, and there are also two bolt holes bored transversely through the piece to receive bolts for reinforcing the end.

A pair of cheekpieces (K), 2" × 4", and each 12 inches long, are mitered at the ends, and each has four bolt holes by means of which the ends may be bolted to the upright (J).

Then a step wedge (L) is made of 1⅜" × 2" material, 10 inches long. This has at least four steps (17), each step being 2 inches long. A wedge 1⅜ inches thick, 10 inches long, and tapering from 2 inches to 1⅜ inches, completes the tail-stock.

The Tool Rest. — This is the most difficult part of the whole lathe, as it must be rigid, and so constructed that it has a revolvable motion as well as being capable of a movement to and from the material in the lathe.

Select a good 4" × 4" scantling (M), 14 inches long, as shown in Fig. 243. Two inches from one end cut a cross gain (I), 1½ inches deep and 1 inch wide, and round off the upper edge, as at 2.

Then prepare a piece (N), 1 inch thick, 8 inches wide, and 10 inches long. Round off the upper edge to form a nose, and midway between its ends [Pg 142] cut a cross gain 4 inches wide and 1½ inches deep. The lower margin may be cut away, at an angle on each side of the gain. All that is necessary now is to make a block (O), 8 inches long, rounded on one edge, and a wedge (P).

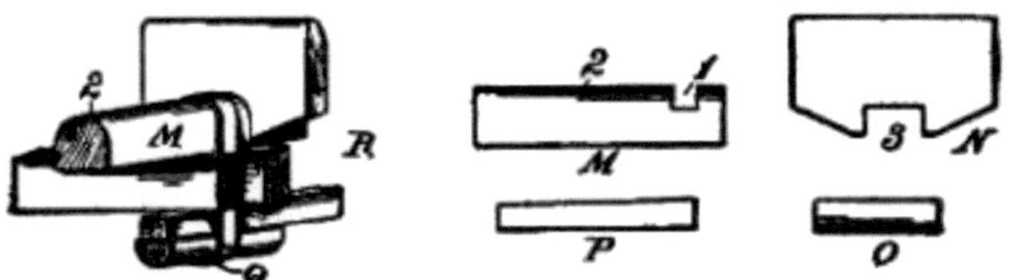

Fig 243. Tool Rest.

A leather belt or strap (Q), 1½ inches wide, formed into a loop, as shown in the perspective view (R), serves as a means for holding the rest rigidly when the wedge is driven in.

The Tool Rest.—This is the most difficult part of the whole lathe, as it must be rigid, and so constructed that it has a revolvable motion as well as being capable of a movement to and from the material in the lathe.

Materials.—Then procure the following bolts:

4⅜"	bolts,	10"	long.
8⅜"	bolts,	6"	long.
20⅜"	bolts,	5"	long.
5⅜"	bolts,	9"	long.

The Mandrel.—A piece of steel tubing (S), No. 10 gage, ¾ inch in diameter, 11½ inches long, will be required for the mandrel. Get a blacksmith, if a machine shop is not convenient, to put a fixed center (1) in one end, and a removable centering member (2) in the other end.

[Pg 143] On this mandrel place a collar (3), held by a set screw, and alongside of it a pair of pulleys, each 1½ inches wide, one of them, being, say, 2 inches in diameter, and the other 3 inches. This mandrel is held in position by means of the posts of the frame which carry the split journal bearings. This form of bearing will make a durable lathe, free from chattering, as the bolts can be used for tightening the mandrel whenever they wear.

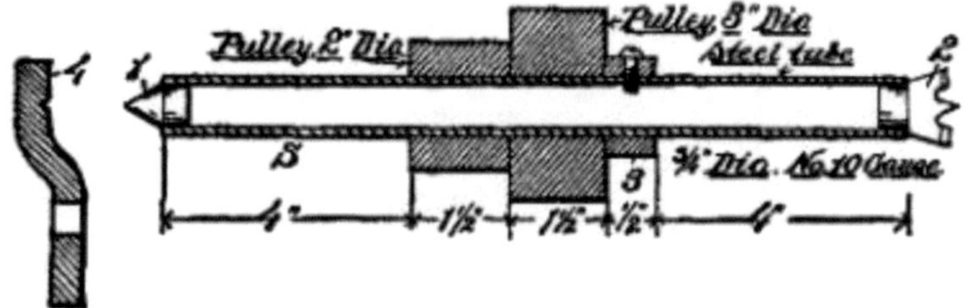

Fig. 244. Mandrel.

The center point (1) is designed to rest against a metal plate (4) bolted to the wooden post, as shown in the large drawing.

Fly-wheel. — It now remains only to provide a fly-wheel and treadle with the communicating belt. The fly-wheel may be of any convenient size, or it may be some discarded pulley or wheel. Suppose it is two feet in diameter; then, as your small pulley is 2 inches in diameter, each revolution of the large wheel makes twelve revolutions in the mandrel, and you can readily turn the wheel eighty [Pg 144] times a minute. In that case your mandrel will revolve 960 revolutions per minute, which is ample speed for your purposes.

The wheel should be mounted on a piece of ¾-inch steel tubing, one end having a crank 3 inches long. This crank is connected up by a pitman rod, with the triangularly shaped treadle frame.

Such a lathe is easily made, as it requires but little metal or machine work, and it is here described because it will be a pleasure for a boy to make such a useful tool. What he needs is the proper plan and the right dimensions to carry out the work, and his own ingenuity will make the modifications suitable to his purpose.

The illustration (Fig. 245) shows such a lathe assembled ready for work.

The Tools Required. — A few simple tools will complete an outfit capable of doing a great variety of work. The illustration (Fig. 246) shows five chisels, of which all other chisels are modifications.

A and B are both oblique firmer chisels, A being ground with a bevel on one side only, and B with a bevel on each side.

C is a broad gage, with a hollow blade, and a curved cutting edge, ground with a taper on the rounded side only.

D is a narrow gage similarly ground, and E is a V-shaped gage.

[Pg 145]

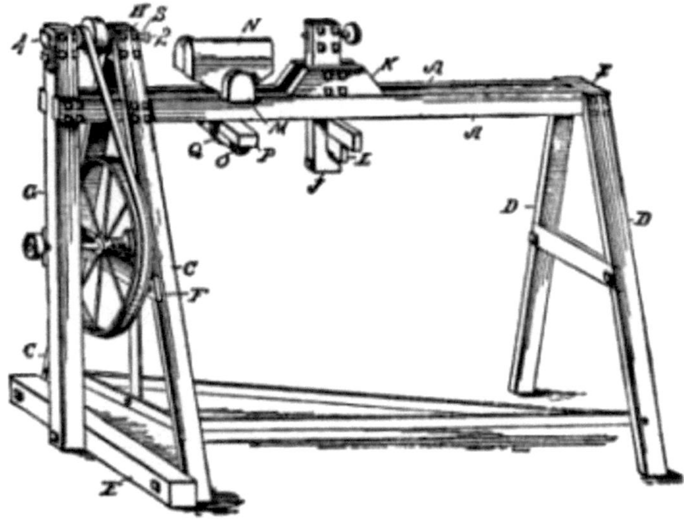

Fig. 245.

[Pg 146]

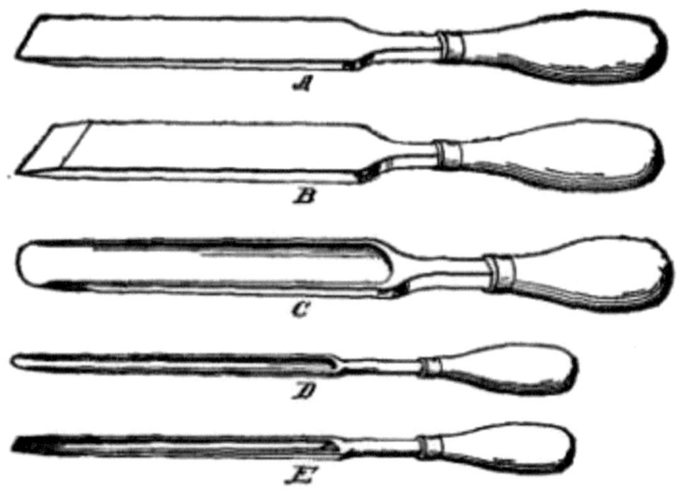

Fig. 246.

It may be observed that in wood-turning sharp tools are absolutely necessary, hence a good oil stone, or several small, round and V-shaped stones should be used.

[Pg 147]

CHAPTER XV

ON THE USE OF STAINS

As this subject properly belongs to the painter and decorator, it is not necessary to go into details concerning the methods used to finish off your work. As you may not be able to afford the luxury of having your productions painted or stained, enough information will be given to enable you, if the character of the wood justifies it, to do the work yourself to a limited extent.

Soft Wood.—As, presumably, most of your first work will be done with pine, poplar, or other light-colored material, and, as many people prefer the furniture to be dark in color, you should be prepared to accommodate them.

Use of Stains.—Our subject has nothing to do with the technique of staining, but has reference, solely, to the use of stains. I recommend, therefore, that, since all kinds of stains are now kept in stock, and for sale everywhere, you would better rely upon the manufactured goods rather than to endeavor to mix up the paints yourself.

Stains as Imitations.—It will be well to remember one thing as to stains. Never attempt to stain anything unless that stain is intended to [Pg 148] produce an imitation of some real wood. There are stains made up which, when applied, do not imitate any known wood. This is bad taste and should be avoided. Again you should know that the same stain tint will not produce like effects on the different light-colored woods. Try the cherry stain on pieces of pine, poplar, and birch, and you will readily see that while pine gives a brilliant red, comparatively speaking, pine or birch will be much darker, and the effect on poplar will be that of a muddy color. In fact, poplar does not stain cherry to good advantage; and for birch the ordinary stain should have a small addition of vermilion.

By making trials of your stains before applying them to the furniture, you will readily see the value of this suggestion.

Good Taste in Staining.—Oak, mahogany, cherry, black walnut, and like imitations are always good in an artistic sense, but imitations of unfamiliar woods mean nothing to the average person. The too common mistake is to try to imitate oak by staining pine or poplar or birch. It may, with good effect, be stained to imitate cherry.

Oregon pine, or some light-colored wood, with a strong contrasting grain may be used for staining in imitation of oak.

Great Contrasts Bad.—Violent contrasts in furniture [Pg 149] staining have the effect of cheapness, unless the contrasting outlines are artistically distributed throughout the article, from base to top finish.

Staining Contrasting Woods.—Then, again, do not stain a piece of furniture so that one part represents a cheap, soft wood, and the other part a dark or costly wood. Imagine, for instance, a cabinet with the stiles, rails and mullions of mahogany, and the panels of pine or poplar, or the reverse, and you can understand how incongruous would be the result produced.

On the other hand, it would not be a very artistic job to make the panels of cherry and the mullions and stiles of mahogany, because the two woods do not harmonize, although frequently wrongly combined.

Hard Wood Imitations.—It would be better to use, for instance, ash or oak for one portion of the work, and a dark wood, like cherry or walnut, for the other part; but usually a cherry cabinet should be made of cherry throughout; while a curly maple chiffonier could not be improved by having the legs of some other material.

These considerations should determine for you whether or not you can safely use stains to represent different woods in the same article.

Natural Effects.—If effects are wanted, the skilled workman will properly rely upon the natural [Pg 150] grain of the wood; hence, in staining, you should try to imitate nature, because in staining you will depend for contrast on the natural grain of the wood to help you out in producing pleasing effects.

Natural Wood Stains.—It should be said, in general, however, that a stain is, at best, a poor makeshift. There is nothing so pleasing as the natural wood. It always has an appearance of cleanliness and openness. To stain the wood shows an attempt to cover up cheapness by a cheap contrivance. The exception to this rule is mahogany, which is generally enriched by the application of a ruby tint which serves principally to emphasize the beautiful markings of the wood.

Polishing Stained Surfaces.—If, on the other hand, you wish to go to the labor of polishing the furniture to a high degree, staining becomes an art, and will add to the beauty and durability of any soft or cheap wood, excepting poplar.

When the article is highly polished, so a good, smooth surface is provided, staining does not cheapen, but, on the other hand, serves to embellish the article.

As a rule, therefore, it is well to inculcate this lesson: Do not stain unless you polish; otherwise, it is far better to preserve the natural color of the wood. One of the most beautiful sideboards I ever [Pg 151] saw was made of Oregon pine, and the natural wood, well filled and highly polished. That finish gave it an effect which enhanced its value to a price which equaled any cherry or mahogany product.

[Pg 152]

CHAPTER XVI

THE CARPENTER AND THE ARCHITECT

A carpenter has a trade; the architect a profession. It is not to be assumed that one vocation is more honorable than the other. A *profession* is defined as a calling, or occupation, "if not mechanical, agricultural, or the like," to which one devotes himself and his energies. A *trade* is defined as an occupation "which a person has learned and engages in, especially mechanical employment, as distinguished from the liberal arts," or the learned professions.

Opportunity is the great boon in life. To the ambitious young man the carpenter's trade offers a field for venturing into the learned professions by a route which cannot be equaled in any other pursuit. In his work he daily enters into contact with problems which require mathematics of the highest order, geometry, the methods of calculating strains and stresses, as well as laying out angles and curves.

This is a trade wherein he must keep in mind many calculations as to materials, number, size, and methods of joining; he must remember all the [Pg 153] small details which go to make up the entire structure. This exercise necessitates a mental picture of the finished product. His imagination is thus directed to concrete objects. As the mind develops, it becomes creative in its character, and the foundation is laid for a higher sphere of usefulness in what is called the professional field.

A good carpenter naturally develops into an architect, and the best architect is he who knows the trade. It is a profession which requires not only the artistic taste, but a technical knowledge of details, of how practically to carry out the work, how to superintend construction, and what the different methods are for doing things.

The architect must have a scientific education, which gives him a knowledge of the strength of materials, and of structural forms; of the durability of materials; of the price, quality, and use of everything which goes into a structure; of labor conditions; and of the laws pertaining to buildings.

Many of these questions will naturally present themselves to the carpenter. They are in the sphere of his employment, but it depends upon himself to make the proper use of the material thus daily brought to him.

It is with a view to instil that desire and ambition in every young man, to make the brain do [Pg 154] what the hand has heretofore done, that I suggest this course. The learned profession is yours if you deserve it, and you can deserve it only through study, application, and perseverance.

Do well that which you attempt to do. *Don't* do it in that manner because some one has done it in that way before you. If, in the trade, the experience of ages has taught the craftsman that some particular way of doing things is correct, there is no law to prevent you from combating that method. Your way may be better. But you must remember that in every plan for doing a thing there is some particular reason, or reasons, why it is carried out in that way. Study and learn to apply those reasons.

So in your leisure or in your active moments, if you wish to advance, you must be alert. *Know for yourself the reasons for things*, and you will thereby form the stepping stones that will lead you upward and contribute to your success.

[Pg 155]

CHAPTER XVII

USEFUL ARTICLES TO MAKE

As stated in the Introductory, the purpose of this book is to show *how to do the things,* and not to draw a picture in order to write a description of it. Merely in the line of suggestion, we give in this chapter views and brief descriptions of useful household articles, all of which may be made by the boy who has carefully studied the preceding pages.

Fig. 247.

This figure shows a common bench wholly made of material 1 inch thick, the top being 12 inches wide and 4 feet long. The legs are 14 inches high and 13 inches wide; and the side supporting rails [Pg 156] are 3 inches wide. These proportions may, of course, be varied. You will note that the sides of the top or seat have an overhang of ½ inch on each margin.

Fig. 248.

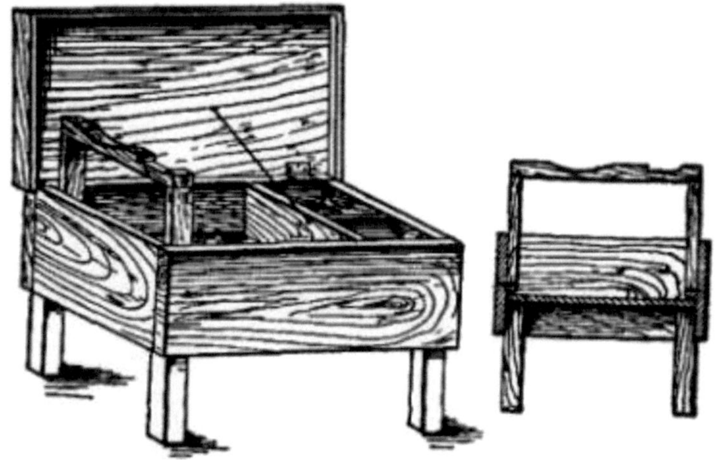

Fig. 249.

This is a common, square-top stool, the seat being 12" × 12", and the legs 14 inches high. Two of the pieces forming the legs are 10 inches wide and the other two 8 inches wide, so that when the [Pg 157] wide pieces are nailed to the edges of the narrow pieces the leg body will be 10" × 10" and thus give the seat an overhang of 1 inch around the margins.

Fig. 250.

A most useful article is shown in Fig. 249. It is a blacking-box with a lid, a folding shoe rest and three compartments. The detached figure shows a vertical cross-section of the body of the box, and illustrates how the shoe rest is hinged to the sides of the box. The box itself is 14" × 16" in dimensions; the sides are 6 inches wide and the legs 5 inches in height. In order to give strength to the legs, the bottom has its corners cut out, to [Pg 158] permit the upper ends of the legs to rest in the recesses thus formed.

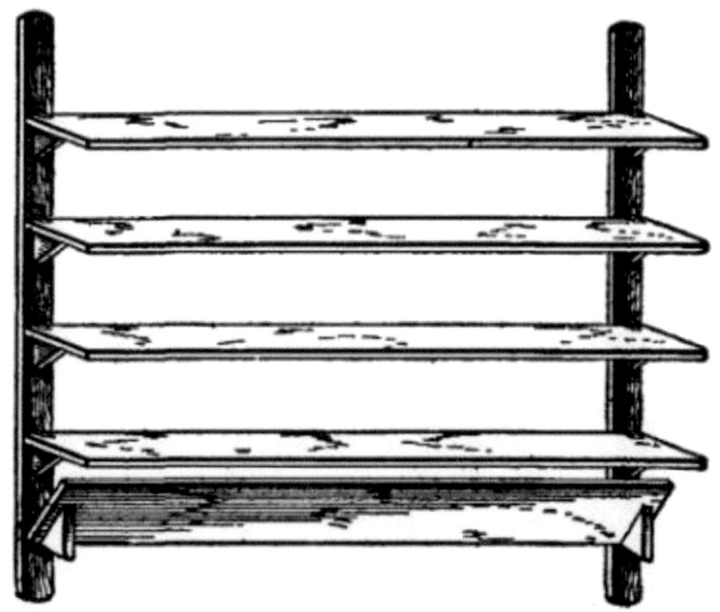

Fig. 251.

This is a convenient form of easel, made of four uprights. The main front uprights are of strips 5/8" × 1¼", and the rear uprights are of ½" × 1" material. A thin broomstick will serve as the pivot bar for the upper end. The rest is made of two strips, each ½" × 1", nailed together to form an L, and nails or wooden pins will serve to hold the rest in any desired position. The front uprights should be at least 5 feet long.

A simple hanging book-rack is illustrated in [Pg 159] Fig. 251. The two vertical strips are each 4 inches wide, 1 inch thick and 4 feet long. Four shelves are provided, each ¾ inch thick, 9 inches wide and 4 feet long. Each shelf is secured to the uprights by hinges on the upper side, so as to permit it to be swung upwardly, or folded; and below each hinge is a triangular block or bracket, fixed to the shelf, to support it in a horizontal position.

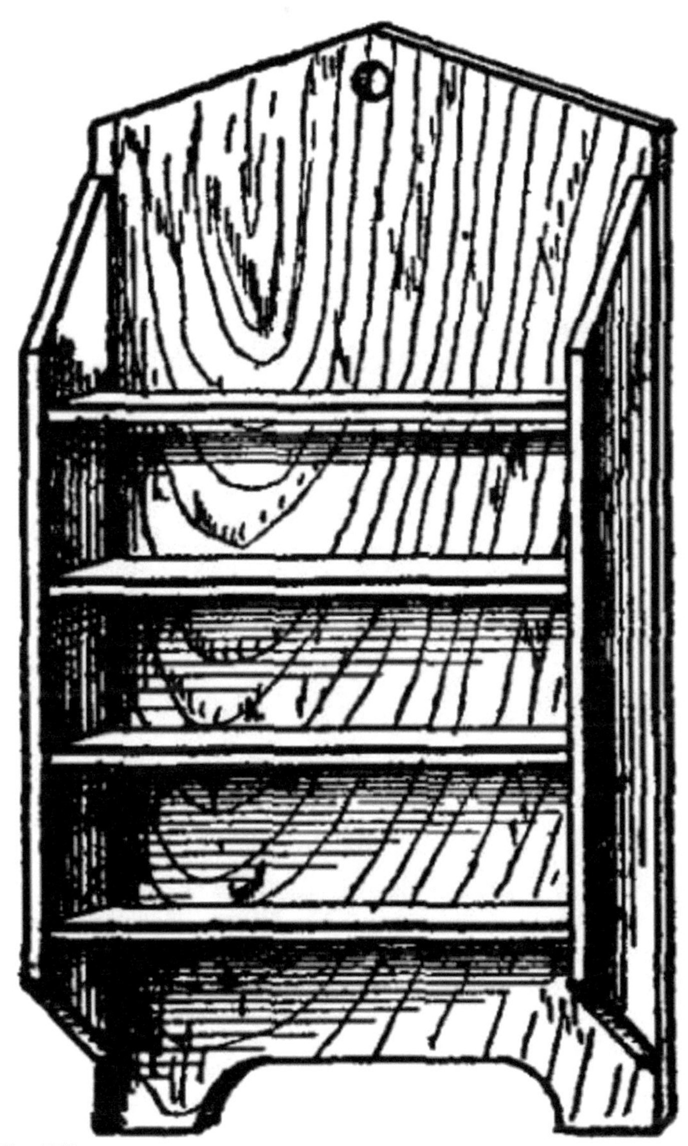

Fig. 252.

A sad-iron holder, or bookcase, shown in Fig. 252, is another simple form of structure. It may be sufficiently large to serve as a standing case by having the uprights at the ends serve as legs, or the uprights may have holes at their upper [Pg 160] ends, by means of which it can be suspended on a wall. As shown, it is 30 inches long from bottom to top, and 20 inches wide. The shelves are 8 inches wide. All the material is, preferably, ¾-inch stock.

Fig. 253.

Fig. 253 shows a wood-box, or it may readily be adapted for coal. For wood it should be 2 feet long, 1 foot 8 inches wide and 1 foot 10 inches high. It will, of course, be made of such dimensions as to suit the wood to be stored in it, and both the flat-top as well as the sloping portion of the top should be hinged, so that the entire top can be opened for filling purposes.

Fig. 254.

Fig. 255.

A pair of parallel bars is shown in Fig. 254. The dimensions of this will vary, and be dependent on the size of the boy intending to use it; but a size best adapted is to make the posts 3 feet high, [Pg 161] and the distance between the bars 16 inches. This gives ample room for the exercises required. The length between the posts along the bars should be at least 5 feet. The entire structure can be made of soft wood, except the bars, which should be of hard, rigid wood. The posts can be made of 2" × 2" material, and the braces 2" × 1".

The base [Pg 162] pieces, both longitudinal and transverse, should also be of 2" × 2" material.

Fig. 256.

Fig. 257.

Fig. 255 represents a mission type of writing desk for a boy's use. All the posts, braces and horizontal bars are of 2" × 2" material, secured to each other by mortises and tenons. The legs [Pg 163] are 27 inches high up to the table top, and the narrow shelf is 12 inches above the top. The most convenient size for the top is 26" × 48". The top boards may be 1 inch thick and the shelf the same thickness, or even ¾ inch. It is well braced and light, and its beauty will depend largely on the material of which it is made.

Fig. 258.

The screen (Fig. 256) represents simply the [Pg 164] framework, showing how simple the structure is. The bars are all of 1½" × 1½" material, secured together by mortises and tenons.

Fig. 257 represents a mission chair to match the desk (Fig. 255), and should be made of the same material. The posts are all of 2" × 2" material. The seat of the chair should be 16 inches, and the rear posts should extend up above the seat at least 18 inches.

Fig. 259.

[Pg 165]

Fig. 260.

Fig. 261.

166

Fig. 258 is a good example of a grandfather's clock in mission style. The framework only is shown. The frame is 12" × 12", and 5 feet high, and made up of 2" × 2" material. When neatly framed together, it is a most attractive article of [Pg 166] furniture. The top may be covered in any suitable way, showing a roof effect. The opening for the dial face of the clock should be at one of the gable ends.

A more pretentious bookcase is shown in Fig. 259, in which the frame is made up wholly of 2" × 2" material. The cross-end bars serve as ledges to support the shelves. This may be lined interiorly and backed with suitable casing material, such as Lincrusta Walton, or fiber-board, and the front provided with doors. Our only object is to show the framework for your guidance, and merely to make suggestions as to structural forms.

Fig. 262.

Another most serviceable article is a case for a coal scuttle (Fig. 260). This should be made of 1-inch boards, and the size of the door, which carries the scuttle shelf, should be 12" × 16" in size. From this you can readily measure the dimensions [Pg 167] of the case itself, the exterior dimensions of which are 15" × 20", so that when the 1-

inch top is placed on, it will be 21 inches high. The case from front to rear is 12 inches, and the shelf above the top is 11 inches wide, and elevated 10 inches above the top of the case. This is a most useful box for culinary articles, if not needed for coal, because the ledge, used for the coal scuttle, can be used to place utensils on, and when the door is opened all the utensils are exposed to view, and are, therefore, much more accessible than if stored away in the case itself.

Fig. 263.

A mission armchair. Fig. 261 is more elaborate than the chair shown in Fig. 257, but it is the same in general character, and is also made of 2" × 2" stock. The seat is elevated 16 inches from the floor, and the rear posts are 28 inches high. [Pg 168] The arms are 8 inches above the seat. A chair of this character should have ample seat space, so the seat is 18" × 18".

The dog house (Fig. 262), made in imitation of a dwelling, is 24 inches square, and 18 inches high to the eaves of the roof. The opening in front is 8" × 10", exclusive of the shaped portion of the opening.

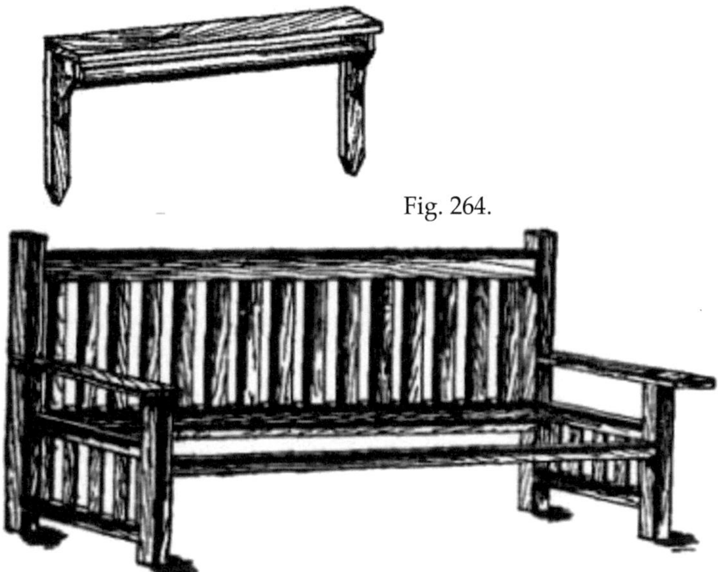

Fig. 264.

Fig. 265.

Fig. 263 shows a simple and easily constructed settee with an under shelf. The seat is 16 inches from the floor and 24 inches wide. The back extends up 24 inches from the seat. The lower shelf is midway between the floor and seat, and [Pg 169] is 19 inches wide. This may or may not be upholstered, dependent on the character of the material of which it is made. If upholstered, the boards may be of second-class material, preferably of pine or other light, soft wood.

A towel rack (Fig. 264) is always a needed article in the kitchen. The roller may be an old curtain roller cut down to 18 inches in length. The top piece is 2½ inches wide and 21 inches long. The vertical bars are each 1½ inches wide and 9 inches long. The brackets are 1½ inches wide and made of ¾-inch material.

Fig. 265 represents the framework of a sofa, the seat of which is 16 inches high, the front posts up to the arm-rests 24 inches, and the rear posts 38 inches. From front to rear the seat is 18 inches. The posts are of 3" × 3" material. This makes a very rigid article of furniture, if mortised and tenoned and properly glued. The seat is 6 feet long, but it may be lengthened or shortened to suit the position in which it is to be placed. It is a companion piece to the chair (Fig. 261).

[Pg 170]

CHAPTER XVIII

SPECIAL TOOLS AND THEIR USES

In the foregoing chapters we have referred the reader to the simple tools, but it is thought desirable to add to the information thus given, an outline of numerous special tools which have been devised and are now on the market.

Bit and Level Adjuster.—It is frequently necessary to bore holes at certain angles. This can be done by using a bevel square, and holding it so one limb will show the boring angle. But this is difficult to do in many cases.

Fig. 266. Bit and Square level.

This tool has three pairs of V slots on its back edges. The shank of the bit will lie in these slots, as shown in Fig. 266, either vertically, or at an angle of 45 degrees, and boring can be done with the utmost accuracy. It may be attached to a Carpenter's square, thus making it an accurate plumb or level.

[Pg 171] Miter Boxes.—The advantages of metal miter boxes is apparent, when accurate work is required.

The illustration, Fig. 267, shows a metal tool of this kind, in which the entire frame is in one solid casting. The saw guide uprights are clamped in tapered sockets in the swivel arm and can be adjusted to hold the saw without play, and this will also counteract a saw that runs out of true, due to improper setting or filing.

Fig. 267. Metal Miter Box.

A second socket in the swivel arm permits the use of a short saw or allows a much longer stroke with a standard or regular saw.

The swivel arm is provided with a tapering index pin which engages in holes placed on the under [Pg 172] side of the base. The edge of the base is graduated in degrees, as plainly shown, and the swivel arm can be set and automatically fastened at any degree desired.

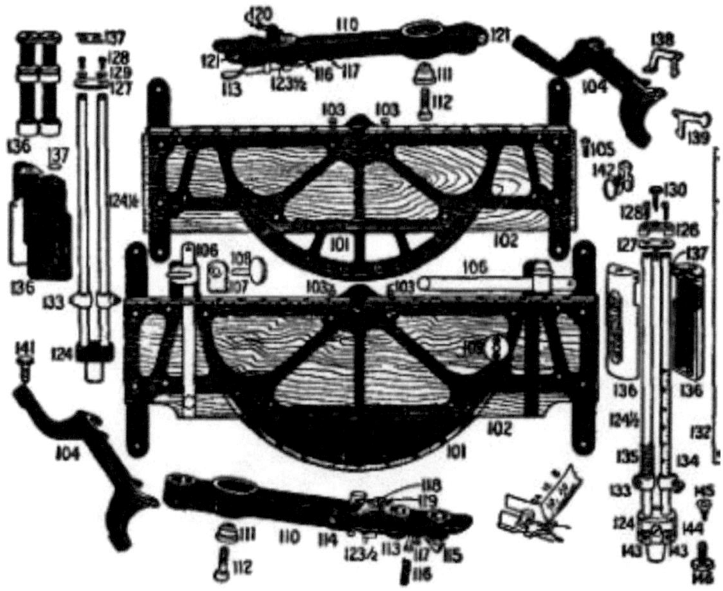

Fig. 268. Parts of Metal Miter Box.

The uprights, front and back are graduated in sixteenths of inches, and movable stops can be set, by means of thumb-screw to the depth of the cut desired.

Figure 268 shows the parts of the miter box, in which the numbers designate the various parts: 101 is the frame; 102 the frame board; 104 frame [Pg 173] leg; 106 guide stock; 107 stock guide clamp; 109 stock guide plate; 110 swivel arm; 111 swivel arm bushing; 112 swivel bushing screw; 113 index clamping lever; 115 index clamping lever catch; 116 index clamping lever spring; 122 swivel complete; 123 T-base; 124½ uprights; 126 saw guide cap; 127 saw guide cap plate; 132 saw guide tie bar; 133 left saw guide stop and screw; 134 right side guide stop and screw; 135 saw guide stop spring; 136 saw guide cylinder; 137 saw guide cylinder plate; 138 trip lever (back); 139 trip lever (front); 141 leveling screw; 142 trip clamp and screw; 146 T-base clamp screw.

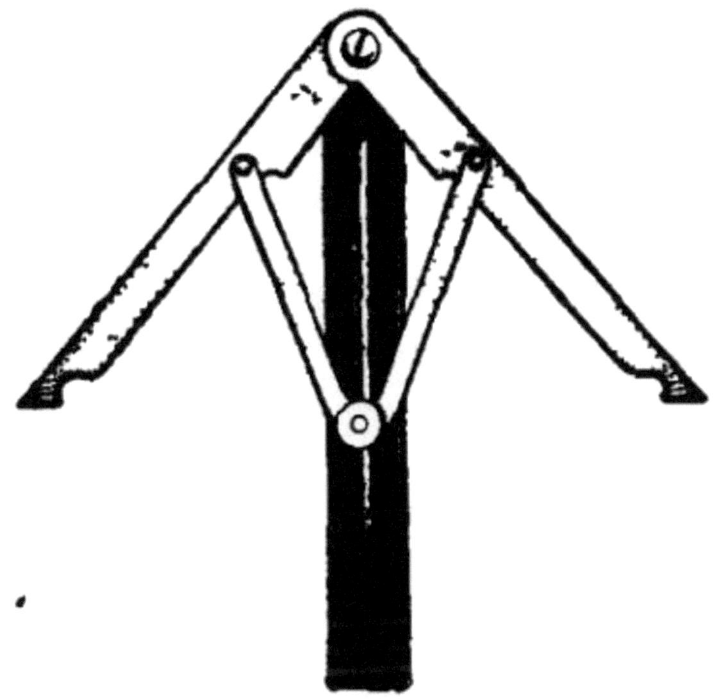

Fig. 269. Angle Dividers.

Angle Dividers. — This is another tool, which does not cost much and is of great service to the [Pg 174] carpenter in fitting moldings where they are applied at odd angles.

To lay out the cut with an ordinary bevel necessitates the use of dividers and a second handling of the bevel, making three operations.

The "Odd Job" Tool. — A most useful special tool, which combines in its make-up a level, plumb try-square, miter-square, bevel, scratch awl, depth gage, marking gage, miter gage, beam compass, and a one-foot rule. To the boy who wishes to economize in the purchase of tools this is an article which should be obtained.

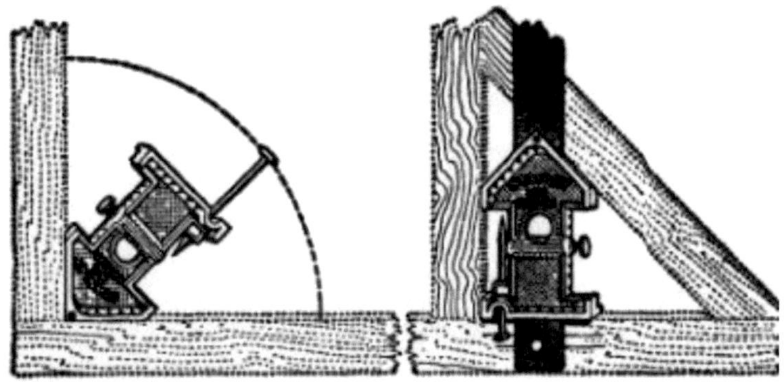

Fig. 270. "Odd Job" Tool.

Figure 270 shows the simplicity of the tool, and how it is applied in use.

Bit Braces. — These tools are now made with so [Pg 175] many improved features that there is really no excuse for getting poor tools.

The illustrations show merely the heads and the lower operating parts of the tools. Fig. 271 shows a metal-clad ball-bearing head, so called, as its under side is completely encased in metal securely screwed to the wood and revolving against the ball thrust bearing.

D represents a concealed ratchet in which the cam ring governs the ratchet, and, being in line with the bit, makes it more convenient in handling than when it is at right angles. The ratchet parts are entirely enclosed, thus keeping out moisture and dirt, retaining lubrication and protecting the users' hands.

The ratchet mechanism is interchangeable, and may be taken apart by removing one screw. The two-piece clutch, which is drop forged, is backed by a very strong spring, insuring a secure lock. When locked, ten teeth are in engagement, while five are employed while working at a ratchet. It has universal jaws (G) for both wood and metal workers.

In Fig. 272, B represents a regular ball bearing head, with the wood screw on the large spindle and three small screws to prevent its working loose. This also has a ball thrust. E is the ratchet box, and this shows the gear teeth cut on the extra [Pg 176] heavy spindle, and encased, so that the user's hands are protected from the teeth.

The interlocking jaws (H), which are best for taper shanks, hold up to No. 2 Clark's expansion, and are therefore particularly adapted for carpenter's use.

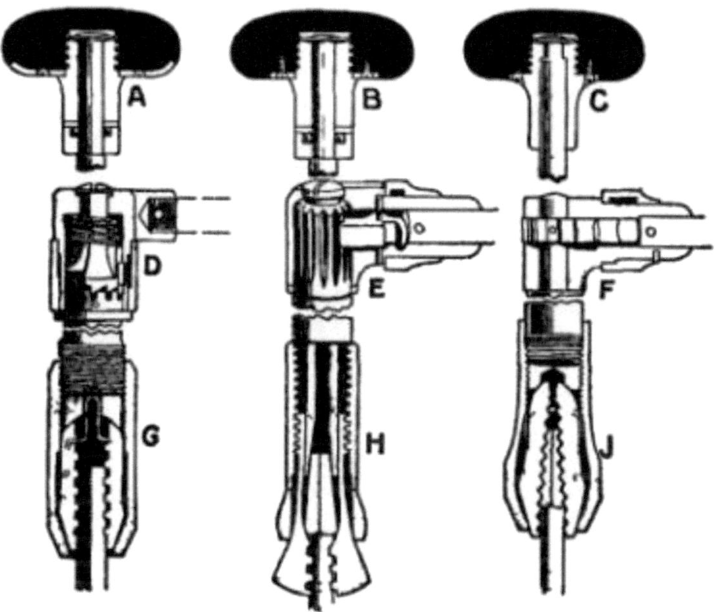

Fig. 271. Fig. 272. Fig. 273.
Types of Bit Braces.

In Fig. 273 the plain bearing head (C) has no ball thrust. The head is screwed on the spindle and [Pg 177] held from turning off by two small screws. The open ratchet (F) shows the gear pinned to the spindle and exposed. This has alligator jaws (J), and will hold all ordinary size taper shank bits, also small and medium round shank bits or drills.

175

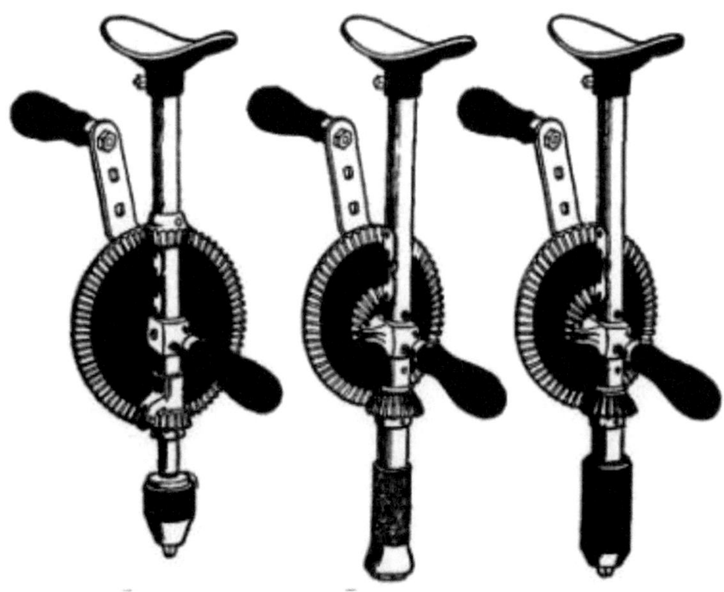

Fig. 274. Fig. 275. Fig. 276.
Steel Frame Breast Drills.

Steel Frame Breast Drill.—These drills are made with both single and double speed, each speed having three varieties of jaws. The single speed is very high, the ratio being 4½ to 1, which makes [Pg 178] it desirable to use for small drills, or for use in wood.

A level is firmly set in the frames of these tools to assist the user to maintain a horizontal position in boring. Each of the forms shown has a ball thrust bearing between the pinion and frame. The breast plate may be adjusted to suit and is locked by a set screw. The spindle is kept from turning while changing drills, by means of the latch mounted on the frame, and readily engaging with the pinion. The crank is pierced in three places so that the handle can be set for three different sweeps, depending on the character of the work.

Figure 274 has a three jaw chuck, and has only single speed. Figure 275 has an interlocking jaw, and is provided with double speed gearing. Figure 276 has a universal jaw, and double speed.

176

Planes.—The most serviceable planes are made in iron, and it might be well to show a few of the most important, to bring out the manner employed to make the adjustments of the bits.

In order to familiarize the boy with the different terms used in a plane, examine Figure 277. The parts are designated as follows: 1A is the double plane iron; 1 single plane iron; 2 plane iron cap; 3 cap screw; 4 lever cap; 5 lever cap screw; 6 frog complete; 7 Y adjusting lever; 8 adjusting nut; 9 lateral adjusting lever; 11 plane handle; [Pg 179] 12 plane knob; 13 handle bolt and nut; 14 knob bolt and nut; 15 plane handle screw; 16 plane bottom; 44 frog pin; 45 frog clamping screw; 46 frog adjusting screw.

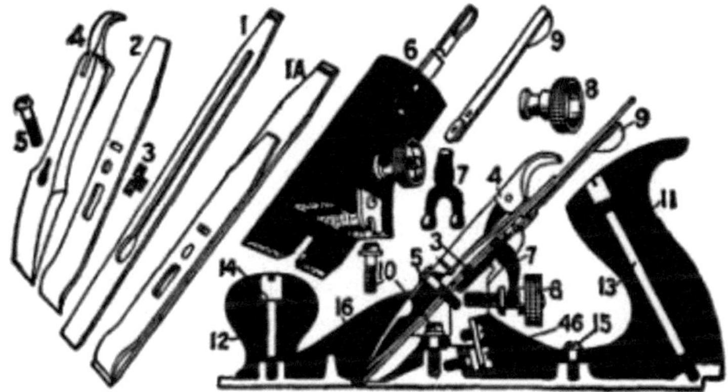

Fig. 277. Details of Metal Plane.

Rabbeting, Matching and Dado Planes.—Figure 278 shows a useful form of plane for the reason that it is designed to receive a variety of irons, adapted to cut rabbets.

The detached sections of Fig. 278 show the various parts, as well as the bits which belong to it. 1, 1 represent the single plane irons; 4 the lever cap; 16 the plane bottom, 50 the fence; 51 the fence thumb screw; 61 the short arm; 70 the adjustable [Pg 180] depth gage; 71 the depth gage which goes through the screw; and 85 the spurs with screws.

Molding and Beading Plane.—A plane of the character shown in Fig. 279 will do an immense variety of work in molding, beading

and dado work, and is equally well adapted for rabbeting, for fil-
letsters and for match planing. The regular equipment with this tool
comprises fifty-two cutters.

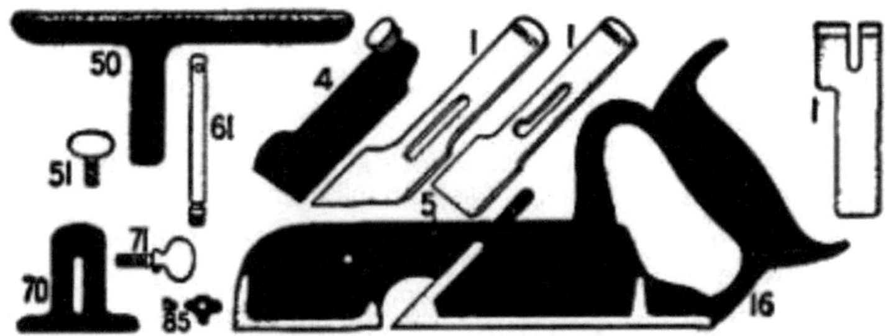

Fig. 278. Rabbet, Matching and Dado Plane.

As shown in Fig. 279, the plane has a main stock (A), which carries the cutter adjustment, a handle, a depth gage, a slitting gage,
and a steel bottom forming a bearing for the other end of the cutter,
and slides on arms secured to the main stock.

This bottom can be raised or lowered, so that, in addition to al-
lowing the use of cutters of different [Pg 181] widths, cutters can be
used having one edge higher or lower than the edge supported in
the main stock.

Fig. 279. Molding and Beading Plane.

The auxiliary center bottom (C), which can be adjusted for width or depth, fulfils the requirement of preventing the plane from tilting and gouging the work. The fence D has a lateral adjustment by means of a screw, for extra fine work. [Pg 182] The four small cuts in the corners show how the bottoms should be set for different forms of cutters, and the great importance of having the fences adjusted so that the cutters will not run.

The samples of work illustrated show some of the moldings which can be turned out with the plane.

Fig. 280. Dovetail Tongue and Groove Plane.

Dovetail Tongue and Groove Plane.—This is a very novel tool, and has many features to recommend it. Figure 280 shows its form, and how it is used. It is designed to make the dovetailed tongue as well as the groove.

It will cut any size groove and tongues to fit with sides of twenty degrees flare, where the width [Pg 183] of the neck is more than one-quarter of an inch thick, and the depth of the groove not more than three-quarters of an inch. The tongue and groove are cut separately, and can be made with parallel or tapering sides. The operation of the plane is very simple.

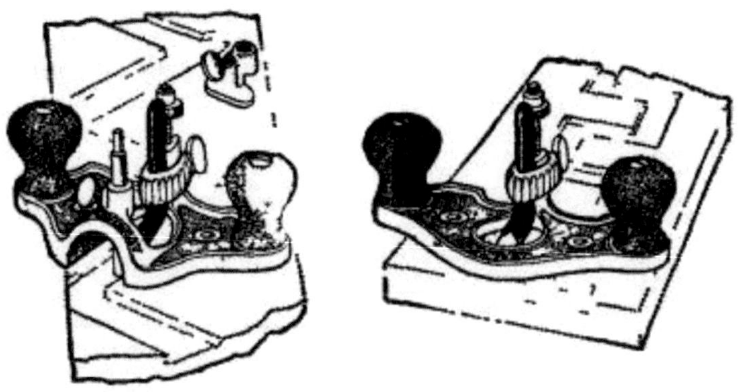

Router Planes.

Router Planes.—This is a type of plane used for surfacing the bottom of grooves or other depressions parallel with the general surface of the work.

The planes are made in two types, one, like Fig. 281, which has a closed throat, and the other, Fig. 282, with an open throat. Both are serviceable, but the latter is preferable. These planes will [Pg 184] level off bottoms of depression, very accurately, and the tool is not an expensive one.

Door Trim Plane.—This is a tool for making mortises for butts, face plates, strike plates, escutcheons, and the like, up to a depth of 5/16, and a width of 3 inches. The principal feature in the plane is the method of mounting the cutter, which can be instantly set to work from either end of the plane or across it.

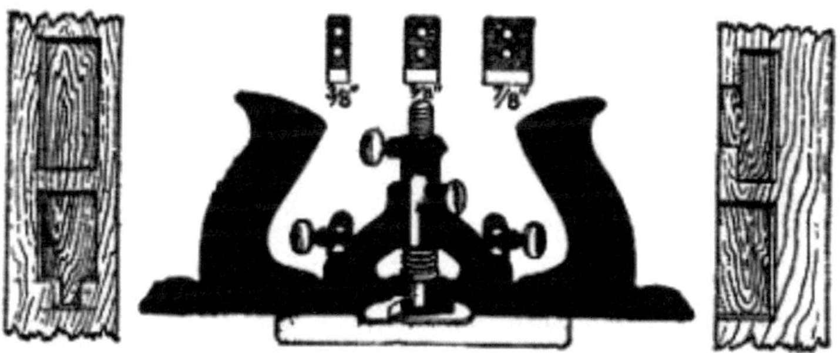

Fig. 283. Door Trim Plane.

The cutter, as shown in Fig. 283, is cushioned by a spring which prevents taking a heavier chip than can be easily carried. A fence regulates the position of the cut and insures the sides of the cut being parallel. The depth of the cut is governed by a positive stop. By removing the fence and locking the cutter post with the thumb screw, instead of using the spring, a very superior router plane is obtained.

[Pg 185]

CHAPTER XIX

ROOFING TRUSSES

The chapter on Bridge Building gives some suggestions as to form of trusses, the particular types there shown being principally for wide spans. Such trusses were made for one purpose only, namely, to take great weight, and they were, as a consequence, so constructed as to provide strength.

But a roofing truss, while designed to hold the accumulated materials, such as snow and ice, likely to be deposited there, is of such a design, principally, so as to afford means of ornamentation. This remark has reference to such types as dispense with the cross, or tie beam, which is the distinguishing feature in bridge building.

The tie beam is also an important element in many types of trusses, where ornamentation is not required, or in such structures as have the roofed portion of the buildings enclosed by ceiling walls, or where the space between the roofs is used for storage purposes.

In England, and on the Continent of Europe, are thousands of trusses structured to support the roofs, which are marvels of beauty. Some of them [Pg 186] are bewildering in their formation. The moldings, beaded surfaces, and the carved outlines of the soffits, of the arches, and of the purlins, are wonderful in detail.

The wooden roof of Westminster Hall, while very simple in structure, as compared with many others, looks like an intricate maze of beams, struts and braces, but it is, nevertheless, so harmonized that the effect is most pleasing to the eye, and its very appearance gives the impression of grandeur and strength.

Nearly all of the forms shown herein have come down to us from mediæval times, when more stress was laid on wooden structures than at the present time, but most of the stone and metal buildings grew out of the wooden prototypes.

Now the prime object of nearly all the double-roofed trusses was to utilize the space between the rafters so as to give height and majesty to the interior.

A large dome is grand, owing to its great simplicity, but the same plain outlines, or lack of ornamentation, in the ceiling of a square or rectangular building would be painful to view, hence, the braces, beams, plates, and various supports of the roofed truss served as ornamental parts, and it is in this particular that the art of the designer finds his inspiration.

[Pg 187] Before proceeding to apply the matter of ornamentation, it might be well to develop these roof forms, starting with the old type Barn Roof, where the space between the rafters must be utilized for the storage of hay.

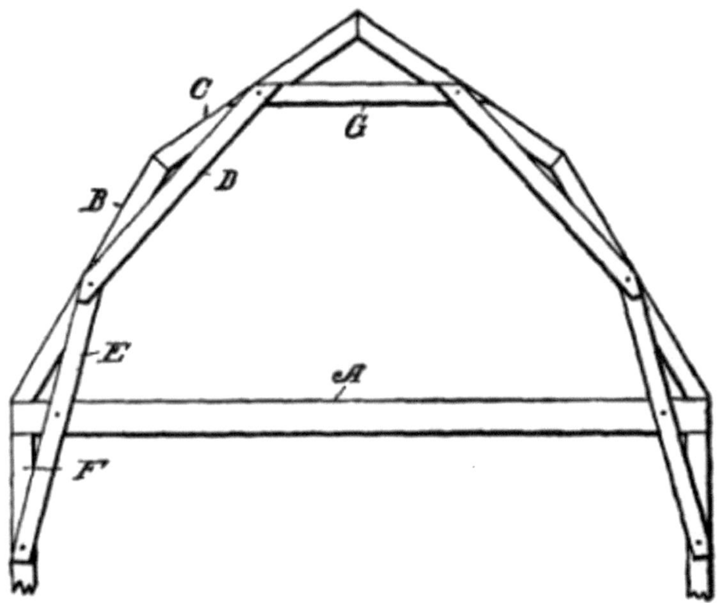

Fig. 284. Gambrel Roof.

The Gambrel Roof, Fig. 284, requires a tie beam, (A), as shown, but the space above the beam is free of all obstructions, and gives a large storage space. The roof has two sets of rafters (B, C), and of different pitch, the lower rafters (B) having a pitch of about 30 degrees, and the upper ones (C), about 45 degrees.

[Pg 188] A tie bar (D) joins the middle portion of each of the rafters (B, C) and another tie bar (E) joins the middle part of the rafter

(B), and the supporting post (F). The cross tie beam (G) completes the span, and a little study will show the complete interdependence of one piece upon the other.

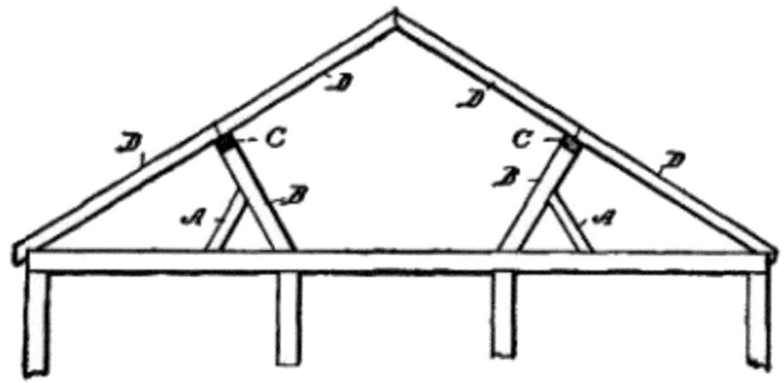

Fig. 285. Purlin Roof.

The Purlin Roof is a type of structure used very largely throughout the United States, for wide barns. (A) is the cross beam; (B, B) the purlin posts; (C, C) the purlin plates; (D, D) the rafters; and (E, E) the supporting braces.

The rafters (D) are in two sections, the distance from the eaves to the comb being too great for single length rafters, and the purlin plates are not designed to make what is called a "self-supporting" roof, but merely to serve as supports for the regular rafters.

[Pg 189] *The Princess Truss*, on the other hand, is designed to act as a support for the different lengths of rafters (A, B, C), and as a means for holding the roof. It is adapted for low pitch and wide spans.

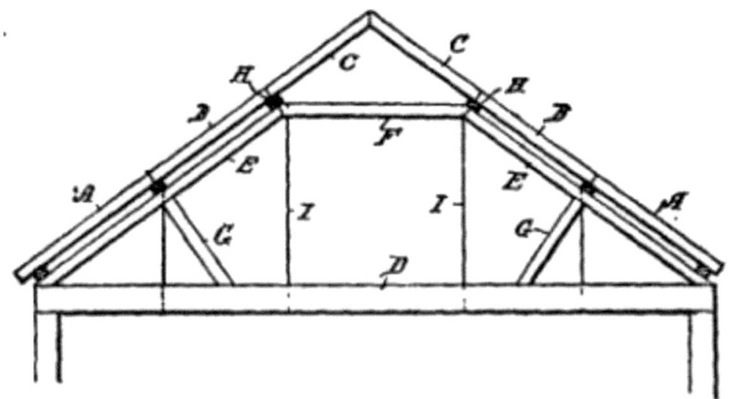

Fig. 286. Princess Truss.

The main truss is made up of the cross beam (D), rafters (E, E) and thrust beam (F). Purlin posts (G, G) are placed at an angle intermediate the ends of the rafters, and the purlin plates (H, H) support the roof rafters (A, B, C); I, I are the vertical tie rods.

This type is probably the oldest form of truss for building purposes, and it has been modified in many ways, the most usual modification being the substitution of posts for the tie rods (I, I).

Following out the foregoing forms, we may [Pg 190] call attention to one more type which permitted ornamentation to a considerable degree, although it still required the tie beam. In fact the tie beam itself was the feature on which the architect depended to make the greatest effect by elaborating it.

This is shown in Fig. 287, and is called the *Arched,* or *Cambered, Tie Beam Truss.* It is a very old type, samples of which have been found which take it back to a very remote age.

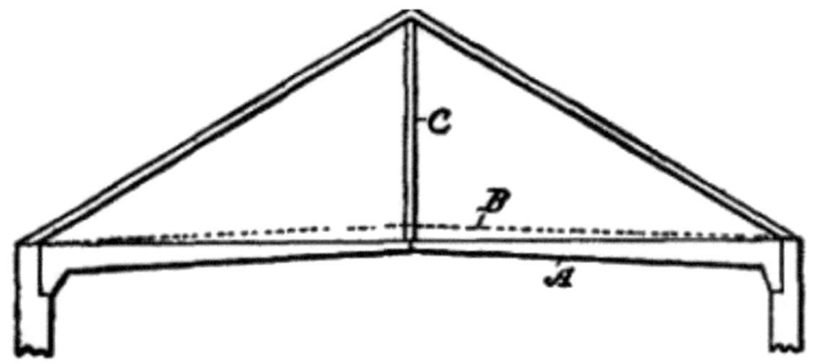

Fig. 287. Arched, or Cambered, Tie Beam.

The tie beam A, in wide spans, was made in two sections, proper-ly tied together, and sometimes the outer ends were very wide, and to add to the effect of the arch, it might also be raised in the middle, something in the form shown by the dotted line (B).

The Mansard is what may be called a double-mounted roof, and it will be seen how it was [Pg 191] evolved from the preceding types. It will be noted that the simple truss formed by the members (A, B, C) is merely superposed on the leaning posts, the tie beam also be-ing necessary in this construction.

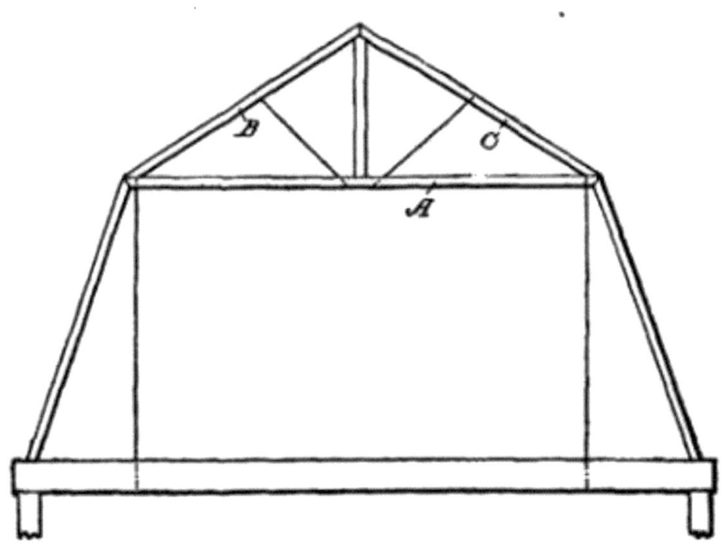

Fig. 288. The Mansard.

But the most elaborate formations are those which were intended to provide trusses for buildings wherein the tie beams were dispensed with.

The simplest form known is called the *Scissors Beam*, illustrated in Fig. 289. This has been utilized for small spaces, and steep pitches. Each rafter (A) has an angled beam or brace (B), springing from its base, to the opposite rafter (A), [Pg 192] to which it is joined, midway between its ends, as at C.

Where the two braces (B) cross each other they are secured together, as at D. As a result, three trusses are formed, namely, 1, 2, 3, and it possesses remarkable strength.

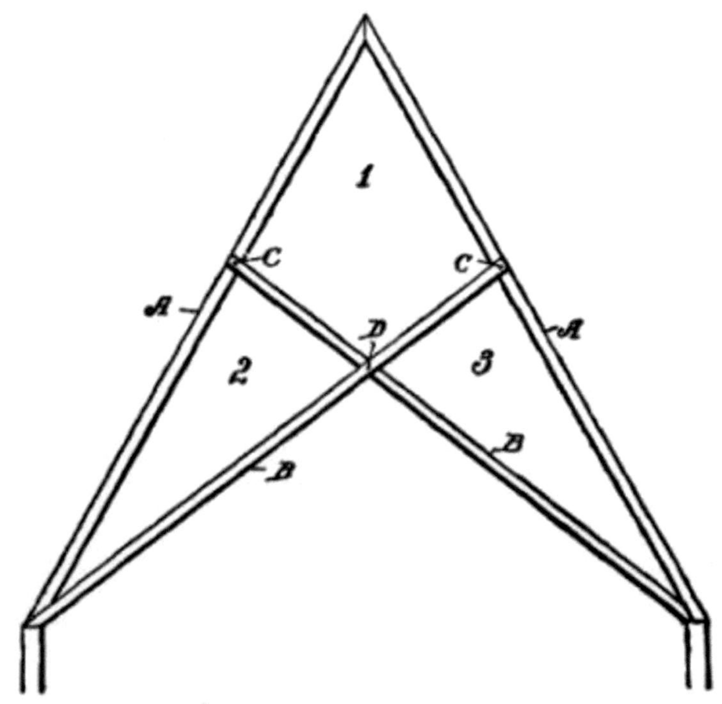

Fig. 289. Scissors Beam.

Braced Collar Beam. — This is a modification of the last type, but is adapted for thick walls only. The tie rod braces (A, A) have to be brought down low to give a good bracing action, and this [Pg 193] arrangement is capable of considerable ornamentation.

The steeper the pitch the higher up would be the inner and lower brace posts (B, B) which were supported by the top of the wall. This form is not available for wide spans, and is shown to illustrate how the development was made into the succeeding types.

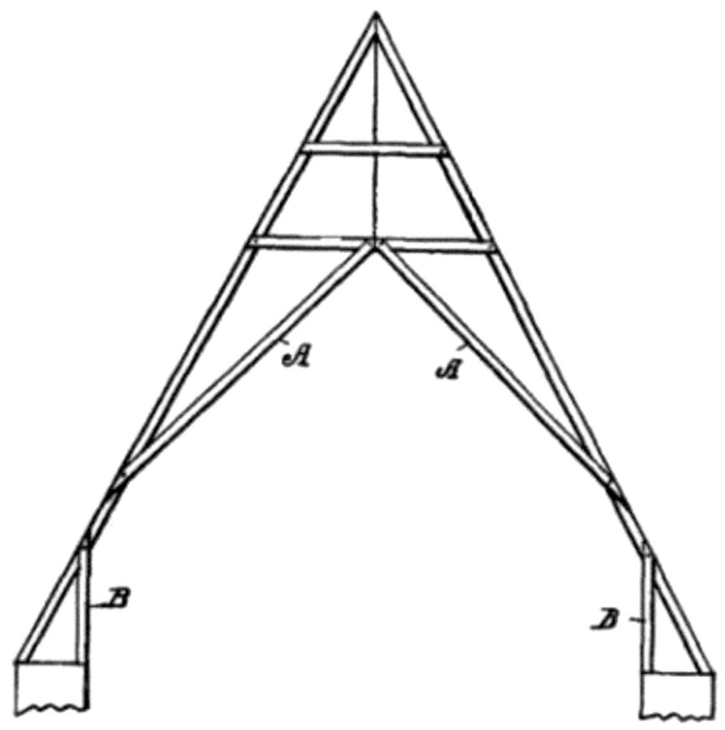

Fig. 290. Braced Collar Beam.

The Rib and Collar Truss, Fig. 291, is the first [Pg 194] important structural arrangement which permitted the architect to give full sway to embellishment. The inwardly-projecting members (A, A) are called *Hammer Beams*. They were devised as a substitute for the thick walls used in the Braced Collar Beam Truss, and small brackets (B, B) were placed beneath as supports.

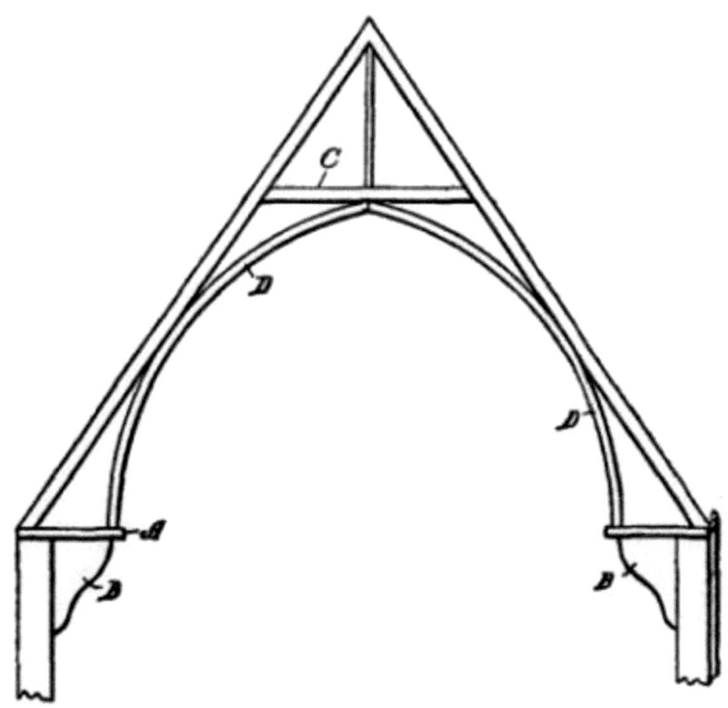

Fig. 291. Rib and Collar Truss.

The short tie beam (C), near the apex, serves as the member to receive the thrust and stress of the curved ribs (D, D). It forms a most graceful type of roof, and is capable of the most exquisite ornamentation, but it is used for the high pitched roofs only.

[Pg 195]

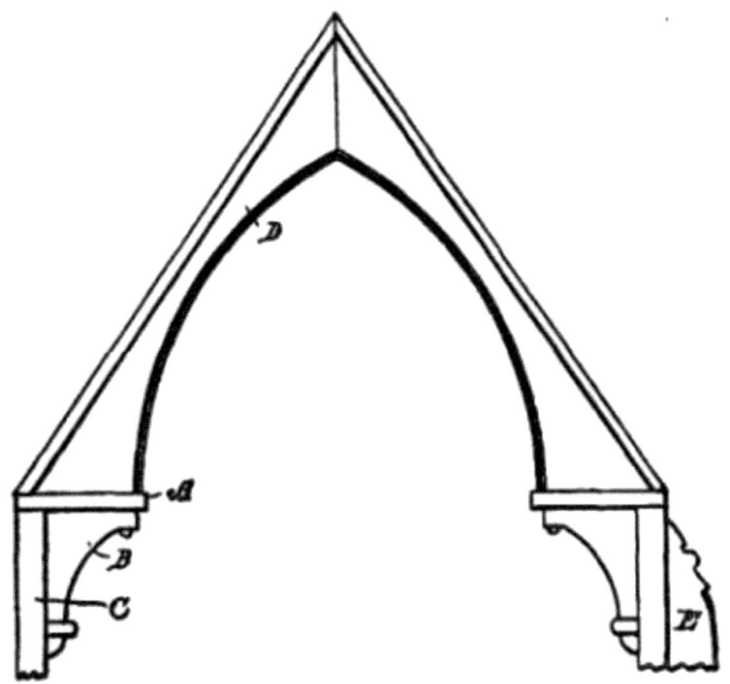

Fig. 291½. Hammer Beam Truss.

The acme of all constructions, in which strength, beauty, and capacity for ornamentation are blended, is the *Hammer Beam Truss*. Here the hammer beam projects inwardly farther than in the preceding figure, and has a deeper bracket (B), and this also extends down the pendant post (C) a greater distance.

[Pg 196] The curved supporting arch (D), on each side, is not ribbed, as in the Rib and Collar Truss, but instead, is provided with openwork (not shown herein), together with beadings and moldings, and other ornamental characteristics, and some of the most beautiful architectural forms in existence are in this type of roof.

What are called Flying Buttresses (E) are sometimes used in connection with the Hammer Beam Truss, which, with heavy roofs and wide spans, is found to be absolutely necessary.

[Pg 197]

CHAPTER XX

ON THE CONSTRUCTION OF JOINTS

192

In uniting two or more elements, some particular type of joint is necessary. In framing timbers, in making braces, in roof construction and supports, in floor beams, and in numerous other places, where strength is required, the workman should have at his command a knowledge of the most serviceable methods.

Illustrations can most forcibly convey the different types; but the sizes must be determined by the character of the material you are working with. Our aim is to give the idea involved, and the name by which each is known.

Fig. 292. Bridle Joints.

Reference has been made in Chapter X, to certain forms of scarfing and lapping pieces. This chapter has to do with a variety of other structural [Pg 198] forms, but principally with such as are used in heavy building work, and in cases where neither fish plates nor scarfing will answer the purpose.

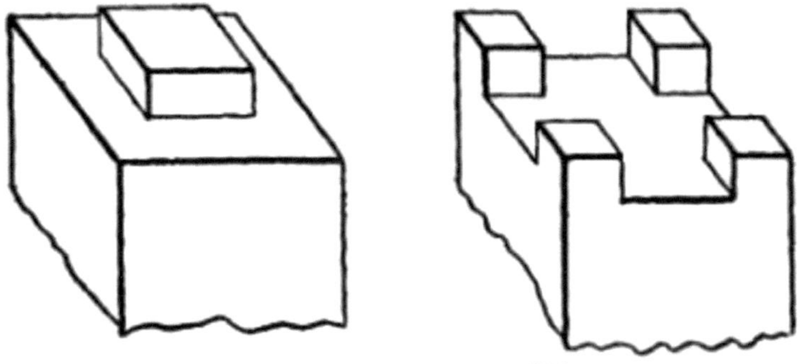

Fig. 293. Spur Tenon.

Fig. 294. Saddle Joints.

Bridle Joints.—This is a form of joint where permanency is not desired, and where it is necessary to readily seat or unseat the vertical timber. It is also obvious that the socket for the upright is of such a character that it will not weaken it to any great extent.

Spur Tenon.—This tenon can be used in many places where the regular one is not available. This, like the preceding, is used where the parts [Pg 199] are desired to be detachable, and the second form is one which is used in many structures.

Saddle Joint.—This is still another manner in which a quickly detachable joint can be constructed. The saddle may be mounted on the main base, or cut into the base piece. An infinite variety of forms of saddles are made, most of them being used in dock work, and for framing of that character where large timbers are used, as in the building of coal chutes, and the like.

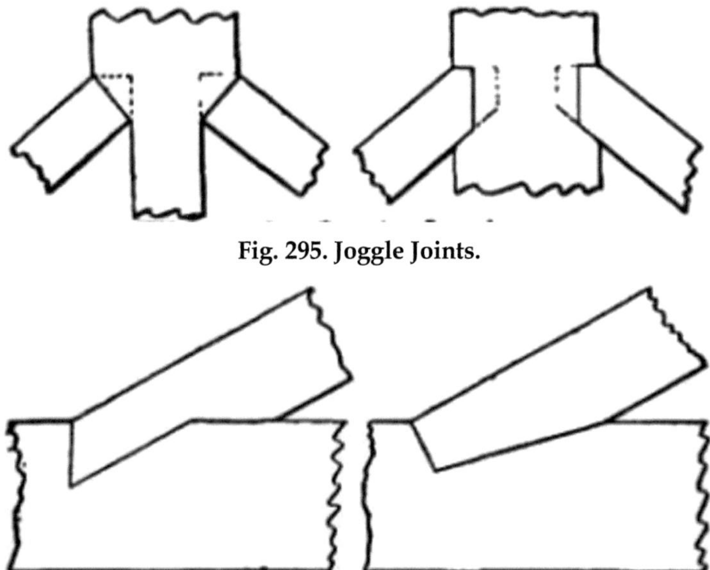

Fig. 295. Joggle Joints.

Fig. 296. Framing Joints.

Joggle Joint.—This joint is used almost exclusively for brace work where great weight must be supported. The brace has a tenon, and the end [Pg 200] must also be so arranged that it will have a direct bearing against the upright, which it braces and supports, or it may have two faces, as in the second figure, which is an exceedingly strong construction.

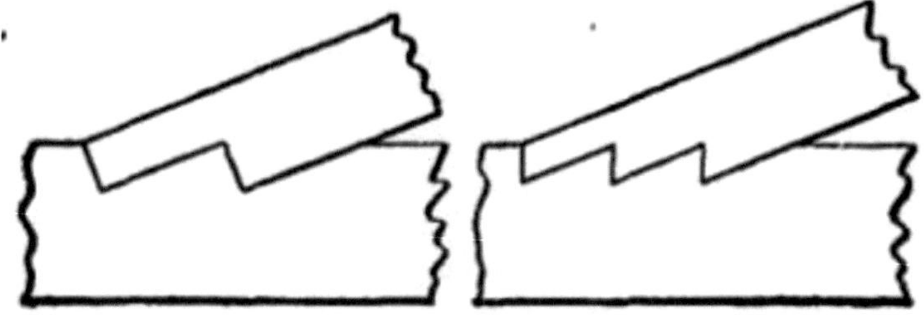

Fig. 297. Heel Joints.

194

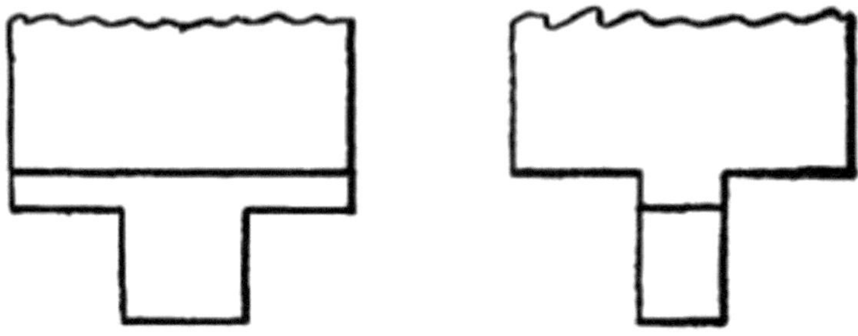

Fig. 298. Stub Tenon.

Framing Joints.—These are the simplest form in which two members are secured together. They are used almost wholly in rafter work, and have very few modifications. The depth of the cut, for the toe of the rafter, depends on the load to be carried, and also on the distance the end of the rafter is from the end of the horizontal member on which the rafter rests.

[Pg 201] Heel Joints.—This is by far the most secure of the framing type of joints. This, if properly made, is much better than the construction shown in the previous illustration, but the difficulty is to make the rafter fit into the recesses properly. This is no excuse for failure to use, but it is on account of inability to make close fits that is accountable for lack of use. It will be seen that in case one of the heels rests against the recess, and the others do not, and the pressure is great, there is a liability to tear out the entire joint.

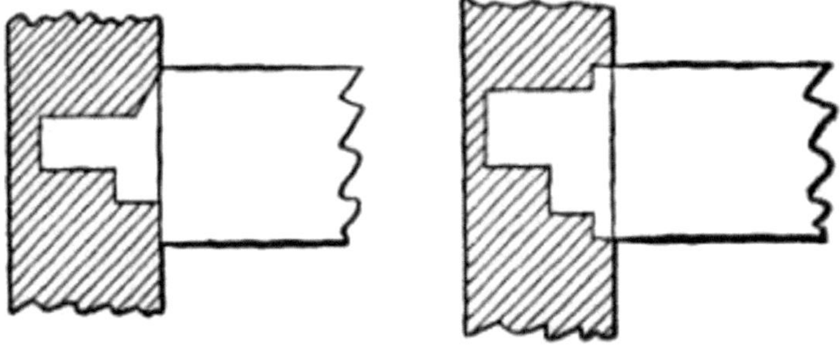

Fig. 299. Tusk Tenon.

Stub Tenon.—This is another form of tenon which is made and designed to be used where it is in close proximity to another tenon, or where the mortises, if made full size, will weaken the member. The long tusk can be shortened, to suit the place where it projects, and the stub tenon on each side of the tusk may be made very short, and one side longer than the other if necessary.

[Pg 202] Tusk Tenon.—Two forms of tusk construction are given. Any number of forms have been devised, all for special purposes, and designed for different kinds of woods. These shown are particularly adapted for soft woods, and the principal feature that is valuable lies in the fact that they have a number of shoulders within the mortise, each of which, necessarily adds to the strength. It should be observed that in the construction of the tusk tenon, the greatest care must be taken to have it fit the mortise tightly, and this has reference to the bottom and shoulder ends as well.

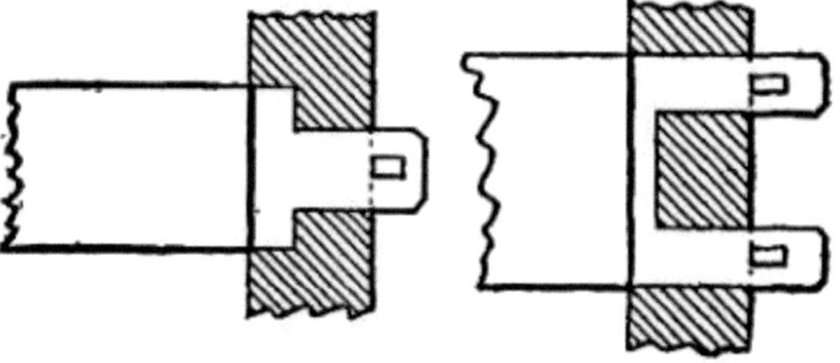

Fig. 300. Double Tusk Tenon.

Double Tusk Tenons.—The distinguishing difference between this and the preceding is in the tusk, which in this form of construction goes through the upright member, and is held by a cross key. The double tusk is intended for hard woods, [Pg 203] and it is regarded as the finest, as well as the strongest, joint known.

Cogged Joints.—This differs from the regular tenoning and mortising methods, principally because the groove or recess is in the form of an open gain. It is used where the member is to be inserted after the main structure is put together.

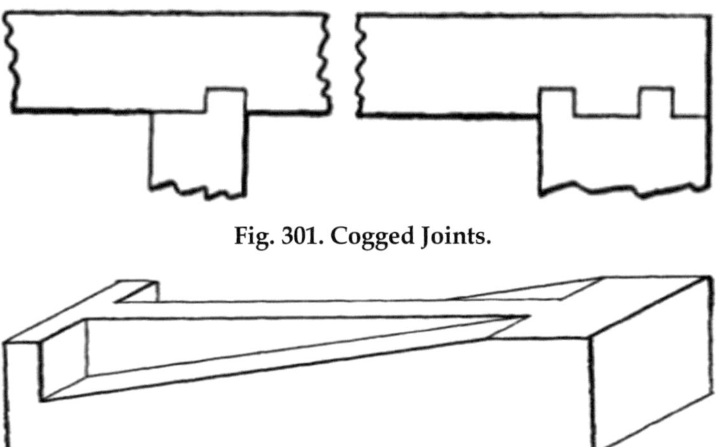

Fig. 301. Cogged Joints.

Fig. 302. Anchor Joint.

Anchor Joint.—This form of connection is designed for very large timbers, and where great care must be taken in making the parts fit together nicely, as everything depends on this. This style [Pg 204] is never used where the angles are less than 45 degrees, and the depth of the gain in the timber receiving the brace is dependent on the thrust of the brace.

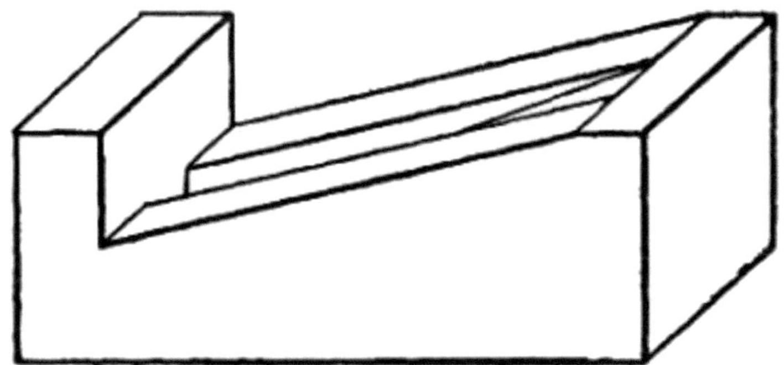

Fig. 303. Deep Anchor Joint.

The Deep Anchor Joint is an extension of the tongue of the Anchor tenon, so that it affords a greater support for the end thrust. To clearly distinguish between this and the preceding form, it might be said that the Anchor Joint is one designed to protect the member containing the gains, while the Deep Anchor Joint favors the brace, by giving it a greater power.

[Pg 205]

CHAPTER XXI

SOME MISTAKES, AND A LITTLE ADVICE IN CARPENTRY

In the mechanical arts, workers are as likely to learn from the mistakes committed as through correct information imparted. Advice, therefore, might be considered superfluous. But there are certain things which are easily remembered and may be borne in mind while engaged in turning out any work.

This chapter is not given for the purpose of calling attention to all the errors which are so common, but merely to point out a few which the boy will commit as he tries to carry out his work for the first time.

One of the difficult things for any one to learn, in working with wood, is to plane the edge of a board straight and square at the same time. This is made doubly difficult if it is desired to plane it strictly to dimensions.

Usually before the edge is straight it is down to the proper width desired, and it is then too late to correct any error, because further work will make it too narrow.

The whole difficulty is in the holding of the plane. It matters not how rigidly it is held, and [Pg 206] how carefully it is guarded to veer it toward one side or the other, it will be found a most difficult task.

If the fore, or finishing, plane is used, and which is the proper tool for the purpose, the impression seems to be, that to square up the edge and make it cut off a thicker shaving on one side than on the other, requires that the plane should be pressed down with force, so as to make it dig in and cut a thicker shaving.

When this is resorted to the board is liable to get out of true from end to end. A much better plan is to put the plane on the edge of the board true and straight. If it is too high on the edge nearest you, bring the plane over so the inside edge is flush with the inside edge of the board.

Then use the fingers of the left hand as a gage to keep the plane from running over.

Now, the weight of the plane in such a condition is sufficient to take off a thicker shaving at the high edge, and this will be done without any effort, and will enable you to concentrate your thoughts on keeping the plane straight with the board.

The weight of the plane will make a thicker shaving on one side than on the other, and correct inequalities, provided you do not attempt to force the plane.

It requires an exceedingly steady hand to hold [Pg 207] a plane firmly for squaring up a half-inch board. Singular as it may seem, it is almost as difficult a job with a two-inch plank. In the case of the thin board the plane will move laterally, unless the utmost care is exercised; in the truing up the thick plank the constant tendency is to move the plane along the surface at a slight diagonal, and this is sure to cause trouble.

It only emphasizes the fact most clearly, that to do a good job the plane must be firmly held, that it must move along the board with the utmost precision, and that it should not be forced into the wood.

In smoothing down a board with the short smoothing plane, preparatory to sandpapering it, the better plan is to move the plane slightly across the grain. This will enable the bit to take hold better, and when the sandpaper is applied the course of the movement should be across the grain opposite the direction taken by the smoothing plane.

It is never satisfactory to draw the sandpaper directly along in the course of the grain. Such a habit will cause the sandpaper to fill up very rapidly, particularly with certain woods.

When gluing together joints or tenons, always wipe off the surplus glue with warm water taken from the glue pot. If you do not follow this advice [Pg 208] the glue will gum up the tools and the sandpaper used to finish the work.

Never try to work from opposite sides of a piece of material. Have a *work side* and a work *edge*, and make all measurements therefrom. Mark each piece as you go along. Take a note mentally just how each piece is to be placed, and what must be done with it.

The carpenter, above all others, must be able to carry a mental picture of his product.

Never saw out the scribing or marking line, either in cutting or in ripping. The lines should be obliterated by the plane, when it is being finished, and not before.

Make it a habit to finish off the surfaces and edges true and smooth before the ends are cut, or the mortises or tenons are made. This is one of the most frequent mistakes. No job can be a perfect one unless your material has been worked down to proper dimensions.

Learn to saw across a board squarely. This may be a hard thing for the novice to do. A long, easy stroke of the saw will prevent it from running, unless too badly set or filed, and will also enable you to hold it more nearly square with the board.

If you find that you invariably saw "out of true," then take some sawing lessons for your own [Pg 209] benefit, until you can judge whether the saw is held true or not.

It is better to saw up a half dozen boards in making the test than commit the error while working on a job.

[Pg 211]

GLOSSARY OF WORDS

USED IN TEXT OF THIS VOLUME

Acute. Sharp, to the point.

Adjuster. A tool which measures distances and relative spaces.

Æsthetic. The theory of taste; science of the beautiful in nature and art.

200

Abstract. That which exists in the mind only; separate from matter; to think of separately as a quality.

Alligator jaws. A term used to designate a pair of serrated bars which are held together in a headpiece, and capable of clamping bits between them.

Analyzed. Separated into its primitive or original parts.

Anchor. Any device for holding an object in a fixed position.

Angle dividers. A sort of double bevel tool so arranged that an angle can be made at the same time on both side of a base line.

Angularly disposed. Forming an angle with reference to some part or position.

Archivolt. The architectural member surrounding the curved opening of an arch. More commonly the molding or other ornaments with which the wall face of an arch is changed.

Artisan. One trained in some mechanic's art or trade.

Beaded. A piece of wood or iron having rounded creases on its surface.

[Pg 212] **Beam compass.** A drawing compass in which the points are arranged to slide on a rod, instead of being fixed on dividers.

Belfry. A bell-tower, usually attached to a church.

Bevel square. A handle to which is pivotally attached a blade, which may be swung and held at any desired angle.

Bisected. To divide, mark, or cut into two portions.

Bit. A small tool, either for drilling, or for cutting, as a plane iron.

Braced collar. A form of roofing truss, in which the upper cross member is supported by a pair of angled braces.

Breast drill. A tool for holding boring tools, and designed to have the head held against the breast for forcing in the boring tool.

Bridle joint. A form for securing elements together which provides a shallow depression in one member, and a chamfered member at its end to fit therein.

Bungalow. A Bengalese term; originally a thatched or tiled house or cottage, single story, usually surrounded by a veranda.

Bushing. A substance of any kind interposed, as, for instance, a wearing surface between a mandrel and its bearing.

Butts. A term applied to certain hinges, usually of the large type.

Callipered. A measured portion which has its side or thickness fixed by a finely graduated instrument.

Cambered. Slightly rising in the middle portion. An upward bend, or projection.

Capital. A small head or top of a column; the head or uppermost member of a pilaster.

[Pg 213] **Cardinal.** Pre-eminent, chief, main line; *Cardinal* line is the principal line to make calculations or measurements from.

Centering point. A place for the reception of the point of an instrument, like a compass or a dividers, or for the dead center of the tail-stock of a lathe.

Cheekpiece. A piece or pieces at right angles to another piece, either fixed or movable, which serves as a rest or a guide.

Chiffonier. A movable and ornamental closet or piece of furniture with shelves and drawers.

Chute. A channel in any material, or made of any substance, for conveying liquids or solids.

Circumference. The distance around an object.

Circumferentially. Surrounding or encircling.

Classical. Relating to the first class or rank, especially in literature or art.

Cogged. Having teeth, either at regular or at irregular intervals.

Concrete. Expressing the thing itself specifically; also the quality; a specific example.

Configuration. Form, as depending on the relative disposition of the parts of a thing; a shape or a figure.

Coincide. To occupy the same place in space; to correspond exactly; to agree; to concur.

Correlation. A reference, as from one thing to another; the putting together of various parts.

Conventional. Something which grows out of or depends upon custom, or is sanctioned by general usage.

Craftsman. One skilled in a craft or trade.

Curvature. The act of curving or being bent.

[Pg 214] **Concentrated.** To bring to a common center; to bring together in one mass.

Dado. A plain flat surface between a base and a surbase molding. Sometimes a painted or encrusted skirting on interior walls.

Depth gage. A tool by means of which the depths of grooves and recesses are measured.

Degree. Measure of advancement; quality; extent; a division or space.

Discarded. Cast off; to reject or put away.

Deterioration. To grow worse; impairing in quality.

Depressed. A sunken surface or part.

Diagrammatical. A drawing made to illustrate the working or the scheme, without showing all the parts or giving their relative positions or measurements.

Diametrically. A direction toward the center or across the middle of a figure or thing.

Diagonal. A direction which is not parallel with or perpendicular to a line.

Dominate. To govern; controlling.

Door trim. The hardware which is attached to a door.

Double-roofed. All form of roof structure where there is an inner frame to support the rafters.

Drop forged. Metal forms which are struck up by means of heavy hammers, in which are the molds or patterns of the article to be formed.

Elaboration. Wrought with labor; finished with great care.

Elevation. The act of raising from a lower to a higher degree; a projection of a building or other object on a plane perpendicular to the horizon.

Elliptical. Having the form of an ellipse.

Embellishment. The act of adorning; that which adds beauty or elegance.

[Pg 215] **Entablature.** The structure which lies horizontally upon the columns.

Equidistant. Being at an equal distance from a point.

Escutcheon. An ornamental plate like that part about a keyhole.

Evolve. To unfold or unroll; to open and expand.

Façade. The front of a building; the principal front having some architectural pretensions.

Facing-boards. The finishing of the face of a wall of different material than the main part of the wall; the wide board below the cornice or beneath the windows.

Factor. One of the elements, circumstances or influences which contribute to produce a result.

Fence. A term used to designate a metal barrier or guard on a part of a tool.

Fish plate. A pair of plates, usually placed on opposite sides of the pieces to be secured together, and held by cross bolts.

Flare. A pitch; an angle; an inclination.

Flush. Unbroken, or even in surface; on a level with the adjacent surface.

Frog clamping screw. A screw which is designed to hold or adjust two angled pieces.

Fulcrum. That by which a lever is sustained, or on which a lever rests in turning or moving a body.

Fluting. The channel or channels in a body; as the grooves in a column.

Gain. A square or beveled notch or groove cut out of a girder, beam, post or other material, at a corner.

Gambrel. A roof having two different pitches, the upper much greater than the lower.

Geometry. Pertaining to that branch of mathematics which investigates [Pg 216] the relations, properties and measurements of solids, surfaces, lines and angles.

Girder. A main beam; a straight horizontal beam to span an opening or carry a weight, such as the ends of floor beams.

Glossary. A collection or explanation of words and passages of the works of an author; a partial dictionary.

Graduated. Cut up into steps; divided into equal parts.

Guide stock. A member which is the main portion of the tool, and from which all measurements are taken.

Hammer beam. A member in a truss roof structure, at the base of the roof proper, which consists of an inwardly projecting part, on which the roof rests, and from which it is braced.

Hammer-pole. The peon, or round end of a hammer which is used for driving nails.

Hemispherical. Pertaining to a half globe or sphere.

Horizontal. On the level; at right angles to a line which points to the center of the earth.

Incorporated. United in one body.

Index pin. A small movable member which is designed to limit the movement of the operative part of a machine.

Initial. To make a beginning with; the first of a series of acts or things.

Insulate. To place in a detached position; to separate from.

Interchangeable. One for the other.

Interval. A space between things; a void space; between two objects.

Interest. To engage the attention of; to awaken or attract attention.

[Pg 217] **Interlocking jaw.** Two or more parts of a piece of mechanism in which the said parts pass each other in their motions.

Intersection. The point or line in which one line or surface cuts another.

Intervening. The portion between.

Inverted. Turned over; to put upside down.

Joggle-joint. A form of connection which has struts attached to a pendant post.

Joinery. The art or trade of joining wood.

Kerf. A notch, channel or slit made in any material by cutting or sawing.

Kit. A working outfit; a collection of tools or implements.

Level. A tool designed to indicate horizontal or vertical surfaces.

Liberal. Not narrow or contracted.

Lobe. Any projection, especially of a rounded form; the projecting part of a cam-wheel.

Longitudinal. In the direction of the length; running lengthwise.

Lubrication. The system of affording oiling means to a machine or to any article.

Mandrel. The live spindle of a lathe; the revolving arbor of a circular saw.

Mansard. A type of roof structure with two pitches, one, the lower, being very steep, and the other very flat pitch.

Manual. Of or pertaining to the hand; done or made by hand.

Marginal. The border or edge of an object.

Marking gage. A bar on which is placed a series of points, usually equidistant from each other.

[Pg 218] **Matching.** Placing tongue in one member and a corresponding groove in another member, so that they will join each other perfectly.

Mediæval. Of or relating to the Middle Ages.

Miter-box. A tool for the purpose of holding a saw true at any desired adjustable angle.

Miter-square. A tool which provides adjustment at any desired angle.

Mullion. A slender bar or pier which forms the vertical division between the lights of windows, screens, etc.; also, indoors, the main uprights are *stiles*, and the intermediate uprights are *mullions*.

Obliterated. Erased or blotted out.

Obtuse. Not pointed; bent.

Orbit. The path made by a heavenly body in its travel around another body.

Ordinate. The distance of any point in a curve or a straight line, measured on a line called the *axis of ordinates,* or on a line parallel to it from another line, at right angles thereto, called the *axis of abscissas*.

Ornamentation. To embellish; to improve in appearance.

Oscillate. To swing like a pendulum.

Overhang. In a general sense that which projects out.

Paneling. A sunken compartment or portion with raised margins, molded or otherwise, as indoors, ceilings wainscoting, etc.

Parallelogram. A right-lined quadrilateral figure, whose opposite sides are parallel and, consequently, equal.

Parallel. Extended in the same direction, and in all parts equally distant.

Perspective. A view; a vista; the effect of distance upon the appearance of objects, by means of which the eye recognizes them as being at a more or less measurable distance.

[Pg 219] **Pivot.** A fixed pin, or short axis, on the end of which a wheel or other body turns.

Pitch. Slope; descent; declivity, like the slope of a roof.

Placement. The act of placing; in the state of being placed.

Predominate. To be superior in number, strength, influence or authority; controlling.

Produced. To lengthen out; to extend.

Prototype. The original; that from which later forms sprang.

Purlin. A longitudinal piece of timber, under a roof, mid-*way between the eaves and comb, to hold the rafters.

Rabbeting. The manner of cutting grooves or recesses.

Ratchet. A wheel, bar, or other form of member, having teeth or recesses.

Rebate. A rectangular, longitudinal recess or groove, cut in the corner or edge of a body.

Rail. A horizontal piece in a frame or paneling.

Rectangular. Right-angled; having one or more angles of ninety degrees; a four-sided figure having only right angles.

Rib and collar. A form of roof truss in which the collar between rafters is used as the thrust bearing for the ribs which project up from the hammer beam.

Router. A tool for cutting grooves or recesses.

Saddle joint. A form of connection in which one part has a portion cut away, resembling a saddle, and in which the part to be attached has its end cut so as to fit the saddle thus formed.

Scarfing. The cutting away of the ends of timbers to be joined, so the two parts on lapping will unite evenly.

Scissors beam. A form of truss, in which there is a pair of interior braces formed like shears, and secured to the main rafters themselves.

[Pg 220] **Score, Scored.** Shear; cut; divide; also notching or marking.

Scratch awl. A sharp-pointed tool, with a handle.

Scribe. To cut, indent or mark with a tool, such as a knife, awl or compass, so as to form a cutting line for the workman.

Self-supporting. Held by itself; not depending upon outside aid.

Shank. Usually the handle, or portion to which the handle is attached.

Slitting gage. A tool which is designed to cut along a certain line guided by an adjustable fence.

Soffit. The under side of an arch.

Solid. Not hollow; full of matter; having a fixed form; hard; opposed to liquid or fluid.

Spindle. A small mandrel; an arbor; a turning shaft.

Springer. The post or point at which an arch rests upon its support, and from which it seems to spring.

Sphere. A body or space continued under a single surface which, in every part, is equally distant from a point within called its center.

Spur. A small part jutting from another.

Strike plate. A plate serving as a keeper for a beveled latch bolt and against which the latter strikes in closing.

Steel Tubing. Pipes made from steel; tubing is measured across from outside to outside; piping is measured on the inside.

Step-wedge. A wedge having one straight edge, and the other edge provided with a succession of steps, by means of which the piece gradually grows wider.

Strain, Stresses. To act upon in any way so as to cause change of form or volume; as forces on a beam to bend it.

[Pg 221] **Strut.** Any piece of timber which runs from one timber to another, and is used to support a part.

Stub. A projecting part, usually of some defined form, and usually designed to enter or engage with a corresponding recess in another member.

Submerged. To be buried or covered, as with a fluid; to put under.

Swivel. A pivoted member, used in many forms of tools, in which one part turns on the other.

Tail-stock. The sliding support or block in a lathe, which carries the dead spindle, or adjustable center.

Technical. Of or pertaining to the useful in mechanical arts, or to any science, business, or the like.

Texture. The disposition of the several parts of any body in connection with each other; or the manner in which the parts are united.

Tool rest. That part of a lathe, or other mechanism, which supports a tool, or holds the tool support.

Torso. The human body as distinguished from the head and limbs.

Transverse. In a crosswise direction; lying across; at right angles to the longitudinal.

Trimmer. A beam, into which are framed the ends of headers in floor framing, as when a hole is left for stairs, chimneys, and the like.

Truss. An assemblage of members of wood or iron, supported at two points, and arranged to transmit pressure vertically to those points with the least possible strain, across the length of any member.

Tusk. In mechanism, a long projecting part, longer than a tenon, and usually applied to the long or projecting part of a tenon.

[Pg 222] **Universal joint.** A joint wherein one member is made to turn with another, although the two turning members are not in a line with each other.

Vocation. Employment; trade; profession; business.

Voissoir. One of the wedgelike stones of which an arch is composed.

THE "HOW-TO-DO-IT" BOOKS

Carpentry for Boys

A book which treats, in a most practical and fascinating manner all subjects pertaining to the "King of Trades"; showing the care and use of tools; drawing; designing, and the laying out of work; the principles involved in the building of various kinds of structures, and the rudiments of architecture. It contains over two hundred and fifty illustrations made especially for this work, and includes also a complete glossary of the technical terms used in the art. The most comprehensive volume on this subject ever published for boys.

Electricity for Boys

The author has adopted the unique plan of setting forth the fundamental principles in each phase of the science, and practically applying the work in the successive stages. It shows how the knowledge has been developed, and the reasons for the various phenomena, without using technical words so as to bring it within the compass of every boy. It has a complete glossary of terms, and is illustrated with two hundred original drawings.

Practical Mechanics for Boys

This book takes the beginner through a comprehensive series of practical shop work, in which the uses of tools, and the structure and handling of shop machinery are set forth; how they are utilized to perform the work, and the manner in which all dimensional work is carried out. Every subject is illustrated, and model building explained. It contains a glossary which comprises a new system of cross references, a feature that will prove a welcome departure in explaining subjects. Fully illustrated.

Price 60 cents per volume

THE NEW YORK BOOK COMPANY

147 Fourth Avenue New York

THE WONDER ISLAND BOYS

By ROGER T. FINLAY

Thrilling adventures by sea and land of two boys and an aged Professor who are cast away on an island with absolutely nothing but their clothing. By gradual and natural stages they succeed in constructing all forms of devices used in the mechanical arts and learn the scientific theories involved in every walk of life. These subjects are all treated in an incidental and natural way in the progress of events, from the most fundamental standpoint without technicalities, and include every department of knowledge. Numerous illustrations accompany the text.

Two Thousand things every boy ought to know. Every page a romance. Every line a fact.

THE NEW YORK BOOK COMPANY

147 Fourth Avenue   New York

The Hickory Ridge Boy Scouts

A SERIES OF BOOKS FOR BOYS

Which, in addition to the interesting boy scout stories by CAPTAIN ALAN DOUGLAS, Scoutmaster, contain articles on nature lore, native animals and a fund of other information pertaining to out-of-door life, that will appeal to the boy's love of the open.

I. The Campfires of the Wolf Patrol

Their first camping experience affords the scouts splendid opportunities to use their recently acquired knowledge in a practical way. Elmer Chenoweth, a lad from the northwest woods, astonishes everyone by his familiarity with camp life. A clean, wholesome story every boy should read.

II. Woodcraft; or, How a Patrol Leader Made Good

This tale presents many stirring situations in which some of the boys are called upon to exercise all their ingenuity and unselfishness. A story filled with healthful excitement.

III. Pathfinder; or, The Missing Tenderfoot

Some mysteries are cleared up in a most unexpected way, greatly to the credit of our young friends. A variety of incidents follow fast, one after the other.

IV. Fast Nine; or, a Challenge From Fairfield

They show the same team-work here as when in camp. The description of the final game with the team of a rival town, and the outcome thereof, form a stirring narrative. One of the best baseball stories of recent years.

V. Great Hike; or, The Pride of The Khaki Troop

After weeks of preparation the scouts start out on their greatest undertaking. Their march takes them far from home, and the good-natured rivalry of the different patrols furnishes many interesting and amusing situations.

VI. Endurance Test; or, How Clear Grit Won the Day

Few stories "get" us more than illustrations of pluck in the face of apparent failure. Our heroes show the stuff they are made of and surprise their most ardent admirers. One of the best stories Captain Douglas has written.

Boy Scout Nature Lore to be Found in The Hickory Ridge Boy Scout Series

Wild Animals of the United States — Tracking — in
Number I.
Trees and Wild Flowers of the United States in
Number II.
Reptiles of the United States in Number III.
Fishes of the United States in Number IV.
Insects of the United States in Number V.
Birds of the United States in Number VI.

Cloth Binding Cover Illustrations in Four Colors 40c. Per Volume

THE NEW YORK BOOK COMPANY

147 FOURTH AVENUE (near 14th St.) NEW YORK

THE

Campfire and Trail Series

1. In Camp on the Big Sunflower.
2. The Rivals of the Trail.
3. The Strange Cabin on Catamount Island.
4. Lost in the Great Dismal Swamp.
5. With Trapper Jim in the North Woods.
6. Caught in a Forest Fire.

By LAWRENCE J. LESLIE

A series of wholesome stories for boys told in an interesting way and appealing to their love of the open.

Each, 12mo. Cloth. 40 cents per volume

THE NEW YORK BOOK COMPANY

**147 FOURTH AVENUE
NEW YORK**

Christy Mathewson's Book

WON IN THE NINTH CHRISTY MATHEWSON

A Ripping Good Baseball Story by One Who Knows the Game

This book has attained a larger sale than any baseball story ever published.

The narrative deals with the students of a large university and their baseball team, the members of which have names which enable the reader to recognize them as some of the foremost baseball stars of the day before their entrance into the major leagues.

One gains a very clear idea of "inside baseball" stripped of wearisome technicalities. The book is profusely illustrated throughout and contains also a number of plates showing the manner in which Mathewson throws his deceptive curves, together with brief description of each.

Cloth bound 5½ x 7⅝ Price 60c. per volume

THE NEW YORK BOOK COMPANY

147 FOURTH AVENUE   NEW YORK

ECONOMICAL COOKING

Primrose Edition

Planned for Two or More Persons

By

MISS WINIFRED S. GIBBS

Dietitian and Teacher of Cooking of the New York Association for Improving the Condition of the Poor

Printed on Fine Quality Book Paper. Cover Design in Colors

Many Cook Books have been published, from time to time, to meet various requirements, or to elucidate certain theories, but very few have been written to meet the needs of the large proportion of our population who are acutely affected by the constantly increasing cost of food products. Notwithstanding that by its valuable suggestions this book helps to reduce the expense of supplying the table, the recipes are so planned that the economies effected thereby are not offset by any lessening in the attractiveness, variety or palatability of the dishes.

Of equal importance are the sections of this work which deal with food values, the treatment of infants and invalids and the proper service of various dishes.

The recipes are planned for two persons, but may readily be adapted for a larger number. The book is replete with illustrations and tables of food compositions—the latter taken from the latest Government statistics.

Cloth Binding Illustrated 40c. per volume, postpaid

THE NEW YORK BOOK COMPANY

147 FOURTH AVENUE (near 14th St.) NEW YORK

CUT-OUT and PAINT BOOKS

SCISSORS BOOK Dolls of All Nations

An original line of art studies printed in full rich colors on high grade paper. This series introduces many novel features of interest, and as the subject matters have been selected with unusual care, the books make a strong appeal not only to the little ones but even to those of riper years.

Post Cards	*Painting Book*
Dolls of all Nations	*Scissors Book*
Our Army	*Scissors Book*
Children's Pets	*Puzzle Book*

Size 8¼ x 10¼ inches

Price 15c. per copy

Send for sample and trade discount

THE NEW YORK BOOK COMPANY

147 FOURTH AVENUE   NEW YORK